Computational Methods in the
Fractional Calculus
of Variations

Computational Methods in the
Fractional Calculus
of Variations

Ricardo Almeida
University of Aveiro, Portugal

Shakoor Pooseh
Technische Universität Dresden, Germany

Delfim F. M. Torres
University of Aveiro, Portugal

Imperial College Press

Published by

Imperial College Press
57 Shelton Street
Covent Garden
London WC2H 9HE

Distributed by

World Scientific Publishing Co. Pte. Ltd.
5 Toh Tuck Link, Singapore 596224
USA office: 27 Warren Street, Suite 401-402, Hackensack, NJ 07601
UK office: 57 Shelton Street, Covent Garden, London WC2H 9HE

Library of Congress Cataloging-in-Publication Data
Almeida, Ricardo (Mathematics professor)
    Computational methods in the fractional calculus of variations / Ricardo Almeida, University of
Aveiro, Portugal, Shakoor Pooseh, Technische Universitat Dresden, Germany, Delfim F.M. Torres,
University of Aveiro, Portugal.
      pages cm
    Includes bibliographical references and index.
    ISBN 978-1-78326-640-1 (hardcover : alk. paper)
    1. Fractional calculus. 2. Calculus of variations. I. Pooseh, Shakoor. II. Torres, Delfim F. M.
III. Title.
    QA314.A46 2015
    515'.83--dc23
                                   2014049076

British Library Cataloguing-in-Publication Data
A catalogue record for this book is available from the British Library.

Printed in Singapore

Preface

The fractional calculus of variations and fractional optimal control are generalizations of the corresponding classical theories, that allow problem modeling and formulations with arbitrary order derivatives and integrals (Malinowska and Torres, 2012). Because of the lack of analytic methods to solve such fractional problems, numerical techniques are developed. In this book, we mainly investigate the approximation of fractional operators by means of series of integer-order derivatives and generalized finite differences. We give upper bounds for the error of proposed approximations and study their efficiency. Direct and indirect methods for solving fractional variational problems are studied in detail. Furthermore, optimality conditions are discussed for different types of unconstrained and constrained variational problems and for fractional optimal control problems. The introduced numerical methods are employed to solve some illustrative examples.

We believe our book provides readers with skillful insights into a fast growing research area, opening new avenues to a world not yet completely explored. It serves as a good complement to (Malinowska and Torres, 2012). In contrast with (Malinowska and Torres, 2012), which deals with the mathematical foundations of fractional variational calculus, this book is devoted to the computational aspects. We trust our readers will find themselves encouraged to participate actively in the exciting area of fractional computational variational methods.

This work is the result of collaboration with many people. We are particularly grateful to all our co-authors: Z. Bartosiewicz, N.R.O. Bastos, M.J. Bohner, A.M.C. Brito da Cruz, J. Cresson, A. Debbouche, M. Dryl, E.H. El Kinani, R.A.C. Ferreira, G.S.F. Frederico, P.D.F. Gouveia, H. Khosravian-Arab, M.J. Lazo, A.B. Malinowska, N. Martins, M.T.T. Monteiro, D. Mozyrska, T. Odzijewicz, A. Plakhov, A. Razminia, H.S. Rodrigues, M. Shamsi, M.R. Sidi Ammi, C.J. Silva, and A. Zinober. We refer the reader to the bibliography at the end.

We are also very obliged to five anonymous referees, for their careful reading of the manuscript and for suggesting invaluable changes, and to the staff at the Imperial College Press, especially to Tasha D'Cruz, Tom Stottor, Jacqueline Downs and Jane Sayers, for taking care of the preparation of the book and for all their support along the process.

A scientific activity is only possible with good financial support, which the Portuguese Foundation for Science and Technology (FCT)—*Fundação para a Ciência e a Tecnologia*—provided us along the years. The work of this book was supported by Portuguese funds through the Center for Research and Development in Mathematics and Applications (CIDMA) and FCT, within project UID/MAT/04106/2013. Torres was also supported by EU funding under the 7th Framework Programme FP7-PEOPLE-2010-ITN, grant agreement 264735-SADCO, and the FCT project PTDC/EEI-AUT/1450/2012, co-financed by FEDER under POFC-QREN with COMPETE reference FCOMP-01-0124-FEDER-028894.

Any comments or suggestions related to the material here contained are more than welcome, and may be submitted by post or by e-mail to the authors. Please contact the authors at:

Ricardo Almeida <ricardo.almeida@ua.pt>
Center for Research and Development in Mathematics and Applications
Department of Mathematics, University of Aveiro
3810-193 Aveiro, Portugal
http://orcid.org/0000-0002-1305-2411

Shakoor Pooseh <Shakoor.Pooseh@tu-dresden.de>
Section of Systems Neuroscience
Department of Psychiatry and Psychotherapy
Faculty of Medicine Carl Gustav Carus
Technische Universität Dresden
Würzburger Straße 35, 01187 Dresden, Germany
http://orcid.org/0000-0002-5441-9507

Delfim F.M. Torres `<delfim@ua.pt>`
Center for Research and Development in Mathematics and Applications
Department of Mathematics, University of Aveiro
3810-193 Aveiro, Portugal
http://orcid.org/0000-0001-8641-2505

Ricardo Almeida, Shakoor Pooseh and Delfim F.M. Torres
January 2015, Aveiro, Portugal

Contents

Preface   v

1. Introduction   1

2. The calculus of variations and optimal control   9

    2.1   The calculus of variations   9
          2.1.1   From light beams to the Brachistochrone problem   9
          2.1.2   Contemporary mathematical formulation   10
          2.1.3   Solution methods   13
    2.2   Optimal control theory   16
          2.2.1   Mathematical formulation   16
          2.2.2   Necessary optimality conditions   18
          2.2.3   Pontryagin's minimum principle   18

3. Fractional calculus   21

    3.1   Special functions   21
    3.2   A historical review   22
    3.3   The relation between the Riemann–Liouville and Caputo
          derivatives   27
    3.4   Integration by parts   28

4. Fractional variational problems   29

    4.1   Fractional calculus of variations and optimal control . . .   29
    4.2   A general formulation   30
    4.3   Fractional Euler–Lagrange equations   32
    4.4   Infinite horizon fractional variational problems   33

	4.4.1	The Euler–Lagrange equation	34
	4.4.2	Optimal control problem	38
	4.4.3	Example	41
4.5		Variational problems with the Riesz–Caputo derivative	42
	4.5.1	The Euler–Lagrange equation	42
	4.5.2	The fractional isoperimetric problem	45
	4.5.3	Optimal time problem	47
4.6		Solution methods	49

5.   Numerical methods for fractional variational problems   51

6.   Approximating fractional derivatives   57

6.1		Riemann–Liouville derivative	57
	6.1.1	Approximation by a sum of integer-order derivatives	57
	6.1.2	Approximation using moments of a function	59
	6.1.3	Numerical evaluation of fractional derivatives	66
	6.1.4	Fractional derivatives of tabular data	68
	6.1.5	Applications to fractional differential equations	72
6.2		Hadamard derivatives	74
	6.2.1	Approximation by a sum of integer-order derivatives	74
	6.2.2	Approximation using moments of a function	76
	6.2.3	Examples	78
6.3		Error analysis	80

7.   Approximating fractional integrals   85

7.1		Riemann–Liouville fractional integral	85
	7.1.1	Approximation by a sum of integer-order derivatives	85
	7.1.2	Approximation using moments of a function	86
	7.1.3	Numerical evaluation of fractional integrals	90
	7.1.4	Applications to fractional integral equations	96
7.2		Hadamard fractional integrals	97
	7.2.1	Approximation by a sum of integer-order derivatives	97
	7.2.2	Approximation using moments of a function	97
	7.2.3	Examples	103

7.3 Error analysis 105

8. Direct methods 111

8.1 Finite differences for fractional derivatives 111
8.2 Euler-like direct method for variational problems 112
    8.2.1 Euler's classic direct method 112
    8.2.2 Euler-like direct method 113
    8.2.3 Examples 116
8.3 A discrete-time method on the first variation 122
    8.3.1 Basic fractional variational problems 125
    8.3.2 An isoperimetric fractional variational problem . . 126

9. Indirect methods 129

9.1 Expansion to integer orders 132
9.2 Expansion through the moments of a function 134

10. Fractional optimal control with free end-points 139

10.1 Necessary optimality conditions 139
     10.1.1 Fractional necessary conditions 140
     10.1.2 Approximated integer-order necessary optimality
            conditions 145
10.2 A generalization 146
10.3 Sufficient optimality conditions 149
10.4 Numerical treatment and examples 151
     10.4.1 Fixed final time 151
     10.4.2 Free final time 155

11. An expansion formula for fractional operators of variable order 159

11.1 Introduction 159
11.2 Fractional calculus of variable order 160
11.3 Expansion formulas with higher-order derivatives 161
11.4 Examples 169
     11.4.1 Test function 169
     11.4.2 Fractional differential equations of variable order . 169
     11.4.3 Fractional variational calculus of variable order . . 172

12. Discrete-time fractional calculus of variations 177

12.1 Introduction 177

12.2   Discrete fractional calculus   178
12.3   Basic definitions on time scales   180
12.4   Calculus of variations on time scales   186
12.5   Fractional variational problems in $\mathbb{T} = \mathbb{Z}$ 189
     12.5.1   Introduction   189
     12.5.2   Preliminaries   190
     12.5.3   Fractional summation by parts   194
     12.5.4   Necessary optimality conditions   197
     12.5.5   Examples   206
     12.5.6   Conclusion   211
12.6   Fractional Variational Problems in $\mathbb{T} = (h\mathbb{Z})_a$ 212
     12.6.1   Introduction   212
     12.6.2   Preliminaries   212
     12.6.3   Fractional h-summation by parts   221
     12.6.4   Necessary optimality conditions   224
     12.6.5   Examples   233
     12.6.6   Conclusion   237

13. Conclusion   239

Appendix A   MATLAB® code   243

Bibliography   251

Index   265

Chapter 1

Introduction

This book is devoted to the study of numerical methods in the calculus of variations and optimal control in the presence of fractional derivatives and/or integrals. A fractional problem of the calculus of variations and optimal control consists in the study of an optimization problem in which the objective functional or constraints depend on derivatives and integrals of arbitrary, real or complex, orders. This is a generalization of the classical theory, where derivatives and integrals can only appear in integer orders. Throughout the book, we call the problems in the calculus of variations and optimal control variational problems. If at least one fractional term exists in the formulation, it is called a fractional variational problem.

The theory started in 1996 with the works of Riewe, in order to better describe nonconservative systems in mechanics (Riewe, 1996, 1997). The subject is now under strong development due to its many applications in physics and engineering, providing more accurate models of physical phenomena (see, e.g., (Almeida, Malinowska and Torres, 2010; Almeida and Torres, 2010; Bastos, Ferreira and Torres, 2011b; Das, 2008; Diethelm, 2010; El-Nabulsi and Torres, 2007, 2008; Ferreira and Torres, 2011; Frederico and Torres, 2007, 2008a; Mozyrska and Torres, 2011; Odzijewicz and Torres, 2011; Ortigueira, 2011; Tenreiro Machado *et al.*, 2010)).

In order to provide a better understanding, the classical theory of the calculus of variations and optimal control is discussed briefly in the beginning of this book, in Chapter 2. Major concepts and notions are presented; key features are pointed out and some solution methods are detailed. There are two major approaches in the classical theory of calculus of variations to solve problems. On the one hand, using Euler–Lagrange necessary optimality conditions, we can reduce a variational problem to the study of a differential equation. Hereafter, one can use either analytical or numerical methods to solve the differential equation and reach the solution of the original problem (see, e.g., (Kirk, 1970)). This approach is referred to as indirect methods in the literature.

On the other hand, we can address the functional itself, directly. Direct methods are used to find the extremizer of a functional in two ways: Euler's finite differences and Ritz methods. In the Ritz method, we either restrict admissible functions to all possible linear combinations

$$x_n(t) = \sum_{i=1}^{n} \alpha_i \phi_i(t),$$

with constant coefficients α_i and a set of known basis functions ϕ_i, or we approximate the admissible functions with such combinations. Using x_n and its derivatives whenever needed, one can transform the functional to a multivariate function of unknown coefficients α_i. By finite differences, however, we consider the admissible functions not on the class of arbitrary curves, but only on polygonal curves made upon a given grid on the time horizon. Using an appropriate discrete approximation of the Lagrangian, and substituting the integral with a sum, and the derivatives by appropriate approximations, we can transform the main problem to the optimization of a function of several parameters: the values of the unknown function on mesh points (see, e.g., (Elsgolts, 1973)).

A historical review of fractional calculus comes next in Chapter 3. In general terms, the field that allows us to define integrals and derivatives of arbitrary real or complex order is called fractional calculus and can be seen as a generalization of ordinary calculus. A fractional derivative of order $\alpha > 0$, when $\alpha = n$ is an integer, coincides with the classical derivative of order $n \in \mathbb{N}$, while a fractional integral is an n-fold integral. The origin of fractional calculus goes back to the end of the seventeenth century, though the main contributions have been made during the last few decades (Sun *et al.*, 2011; Tenreiro Machado, Kiryakova and Mainardi, 2011). Namely it has been proven to be a useful tool in engineering and optimal control problems (see, e.g., (Bhalekar *et al.*, 2012a,b; Efe, 2012; Jarad, Abdeljawad and Baleanu, 2013; Li, Chen and Ahn, 2011; Shen and Cao, 2012)). Furthermore, during the last three decades, several numerical methods have been developed in the field of fractional calculus. Some of their advantages, disadvantages, and improvements, are given in (Aoun *et al.*, 2004).

There are several different definitions of fractional derivatives in the literature, such as Riemann–Liouville, Grünwald–Letnikov, Caputo, etc. They possess different properties: each one of those definitions has its own advantages and disadvantages. Under certain conditions, however, they are equivalent and can be used interchangeably. The Riemann–Liouville and Caputo are the most common for fractional derivatives, and for fractional integrals the usual one is the Riemann–Liouville definition.

After some introductory arguments of classical theories for variational problems and fractional calculus, the next step is providing the framework that is required to include fractional terms in variational problems and is shown in Chapter 4. In this framework, the fractional calculus of variations and optimal control are research areas under strong current development. For the state of the art, we refer the reader to the recent book (Malinowska and Torres, 2012), for models and numerical methods we refer to (Baleanu *et al.*, 2012).

A fractional variational problem consists in finding the extremizer of a functional that depends on fractional derivatives and/or integrals subject to some boundary conditions and possibly some extra constraints. As a simple example one can consider the following minimization problem:

$$J[x] = \int_a^b L(t, x(t), {}_aD_t^\alpha x(t))dt \longrightarrow \min,$$
$$x(a) = x_a, \quad x(b) = x_b,$$

(1.1)

that depends on the left Riemann–Liouville derivative, ${}_aD_t^\alpha$. Although this has been a common formulation for a fractional variational problem, the consistency of fractional operators and the initial conditions is questioned by many authors. For further readings we refer to (Ortigueira and Coito, 2008, 2012; Trigeassou, Maamri and Oustaloup, 2013) and references therein.

An Euler–Lagrange equation for this problem has been derived first in (Riewe, 1996, 1997), see also (Agrawal, 2002). A generalization of the problem to include fractional integrals, the transversality conditions and many other aspects can be found in the literature of recent years. See (Almeida and Torres, 2010; Atanacković, Konjik and Pilipović, 2008; Malinowska and Torres, 2012) and references therein. Indirect methods for fractional variational problems have a vast background in the literature and can be considered a well-studied subject: see (Agrawal, 2002; Almeida, Pooseh and Torres, 2012; Atanacković, Konjik and Pilipović, 2008; Frederico and Torres, 2010; Jelicic and Petrovacki, 2009; Klimek, 2001; Odzijewicz, Malinowska and Torres, 2012b; Riewe, 1997) and references therein that study different variants of the problem and discuss a range of possibilities in the presence of fractional terms, Euler–Lagrange equations and boundary conditions. With respect to results on fractional variational calculus via Caputo operators, we refer the reader to (Agrawal, 2007b; Almeida, Malinowska and Torres, 2012; Almeida and Torres, 2011; Frederico and Torres, 2010; Malinowska and Torres, 2010e; Mozyrska and Torres, 2010; Odzijewicz, Malinowska and Torres, 2012a) and references therein.

Direct methods, however, to the best of our knowledge, have received less interest and are not well studied. A brief introduction of using finite differences has been made in (Riewe, 1996), which can be regarded as a predecessor to what we call here an Euler-like direct method. A generalization of Leitmann's direct method can be found in (Almeida and Torres, 2010), while (Lotfi, Dehghan and Yousefi, 2011) discusses the Ritz direct method for optimal control problems that can easily be reduced to a problem of the calculus of variations.

It is well-known that for most problems involving fractional operators, such as fractional differential equations or fractional variational problems, one cannot provide methods to compute the exact solutions analytically. Therefore, numerical methods are being developed to provide tools for solving such problems. Using the Grünwald–Letnikov approach, it is convenient to approximate the fractional differentiation operator, D^α, by generalized finite differences. In (Podlubny, 1999) some problems have been solved by this approximation. In (Diethelm, Ford and Freed, 2002) a predictor-corrector method is presented that converts an initial value problem into an equivalent Volterra integral equation, while (Kumar and Agrawal, 2006) shows the use of numerical methods to solve such integral equations. A good survey on numerical methods for fractional differential equations can be found in (Ford and Connolly, 2006).

A numerical scheme to solve fractional Lagrange problems has been presented in (Agrawal, 2004). The method is based on approximating the problem to a set of algebraic equations using some basis functions. See Chapter 5 for details. A more general approach can be found in (Tricaud and Chen, 2010a) that uses the Oustaloup recursive approximation of the fractional derivative, and reduces the problem to an integer-order (classical) optimal control problem. A similar approach is presented in (Jelicic and Petrovacki, 2009), using an expansion formula for the left Riemann–Liouville fractional derivative developed in (Atanacković and Stanković, 2004, 2008), to establish a new scheme to solve fractional differential equations.

The scheme is based on an expansion formula for the Riemann–Liouville fractional derivative. Here we introduce a generalized version of this expansion, in Chapter 6, that results in an approximation, for left Riemann–Liouville derivative, of the form

$$_aD_t^\alpha x(t) \approx A(t-a)^{-\alpha}x(t) + B(t-a)^{1-\alpha}\dot{x}(t) - \sum_{p=2}^{N} C(\alpha,p)(t-a)^{1-p-\alpha}V_p(t),$$

(1.2)

with

$$\begin{cases} \dot{V}_p(t) = (1-p)(t-a)^{p-2}x(t) \\ V_p(a) = 0, \end{cases}$$

where $p = 2, \ldots, N$, and the coefficients $A = A(\alpha, N)$, $B(\alpha, N)$ and $C(\alpha, p)$ are real numbers depending on α and N. The number N is the order of approximation. Together with a different expansion formula that has been used to approximate the fractional Euler–Lagrange equation in (Atanacković, Konjik and Pilipović, 2008), we perform an investigation of the advantages and disadvantages of approximating fractional derivatives by these expansions. The approximations transform fractional derivatives into finite sums containing only derivatives of integer order (Pooseh, Almeida and Torres, 2013a).

We show the efficiency of such approximations to evaluate fractional derivatives of a given function in closed form. Moreover, we discuss the possibility of evaluating fractional derivatives of discrete tabular data. The application to fractional differential equations is also developed through some concrete examples.

The same ideas are extended to fractional integrals in Chapter 7. Fractional integrals appear in many different contexts, e.g., when dealing with fractional variational problems or fractional optimal control (Almeida, Malinowska and Torres, 2010; Almeida, Pooseh and Torres, 2012; Frederico and Torres, 2008a; Malinowska and Torres, 2010e; Mozyrska and Torres, 2011). Here we obtain a simple and effective approximation for fractional integrals. We obtain decomposition formulas for the left and right fractional integrals of functions of class C^n (Pooseh, Almeida and Torres, 2012a).

In this book we also consider the Hadamard fractional integral and fractional derivative (Pooseh, Almeida and Torres, 2012c). Although the definitions go back to the works of Hadamard (1892), this type of operators is not yet well studied and much exists to be done. For related works on Hadamard fractional operators, see (Butzer, Kilbas and Trujillo, 2002, 2003; Katugampola, 2011; Kilbas, 2001; Kilbas and Titioura, 2007; Qian, Gong and Li, 2010).

An error analysis is given for each approximation whenever needed. These approximations are studied throughout some concrete examples. In each case we try to analyze problems for which the analytic solution is available, so we can compare the exact and the approximate solutions. This approach gives us the ability of measuring the accuracy of each method. To this end, we need to measure how close we get to the exact solutions. We can use the 2-norm for instance, and define an error function $E[x, \tilde{x}]$ by

$$E = \|x - \tilde{x}\|_2 = \left(\int_a^b [x(t) - \tilde{x}(t)]^2 dt \right)^{\frac{1}{2}}, \qquad (1.3)$$

where x and $\tilde{x}$ are defined on a certain interval $[a, b]$.

Before getting into the usage of these approximations for fractional variational problems, we introduce an Euler-like discrete method, and a discretization of the first variation to solve such problems in Chapter 8. The finite differences approximation for integer-order derivatives is generalized to derivatives of arbitrary order and gives rise to the Grünwald–Letnikov fractional derivative. Given a grid on $[a, b]$ as $a = t_0, t_1, \ldots, t_n = b$, where $t_i = t_0 + ih$ for some $h > 0$, we approximate the left Riemann–Liouville derivative as

$$_aD_{t_i}^\alpha x(t_i) \simeq \frac{1}{h^\alpha} \sum_{k=0}^i (\omega_k^\alpha) \, x(t_i - kh),$$

where $(\omega_k^\alpha) = (-1)^k \binom{\alpha}{k} = \frac{\Gamma(k-\alpha)}{\Gamma(-\alpha)\Gamma(k+1)}$. The method follows the same procedure as in the classical theory. Discretizing the functional by a quadrature rule, integer-order derivatives by finite differences and substituting fractional terms by corresponding generalized finite differences, results in a system of algebraic equations. Finally, one gets approximate values of state and control functions on a set of discrete points (Pooseh, Almeida and Torres, 2013b).

A different direct approach for classical problems has been introduced in (Gregory and Lin, 1993; Gregory and Wang, 1990). It uses the fact that the first variation of a functional must vanish along an extremizer. That is, if x is an extremizer of a given variational functional J, the first variation of J evaluated at x, along any variation η, must vanish. This means that

$$J'[x, \eta] = \int_a^b \left[\frac{\partial L}{\partial x}(t, x(t), \,_aD_t^\alpha x(t))\eta(t) \right.$$
$$\left. + \frac{\partial L}{\partial \,_aD_t^\alpha x}(t, x(t), \,_aD_t^\alpha x(t)) \,_aD_t^\alpha \eta(t) \right] dt = 0.$$

With a discretization on time horizon and a quadrature for this integral, we obtain a system of algebraic equations. The solution to this system gives an approximation to the original problem (Pooseh, Almeida and Torres, 2013e).

Considering indirect methods in Chapter 9, we transform the fractional variational problem into an integer-order problem. The job is done by substituting the fractional term by the corresponding approximation in which only integer-order derivatives exist. The resulting classic problem, which is considered as the approximated problem, can be solved by any available method in the literature. If we substitute the approximation (1.2) for the fractional term in (1.1), the outcome is an integer-order constrained variational problem

$$J[x] \approx \int_a^b L\left(t, x(t), \frac{Ax(t)}{(t-a)^\alpha} + \frac{B\dot{x}(t)}{(t-a)^{\alpha-1}} - \sum_{p=2}^N \frac{C(\alpha,p)V_p(t)}{(t-a)^{p+\alpha-1}}\right) dt$$

$$= \int_a^b L'\left(t, x(t), \dot{x}(t), V_2(t), \ldots, V_N(t)\right) dt \longrightarrow \min$$

subject to

$$\begin{cases} \dot{V}_p(t) = (1-p)(t-a)^{p-2}x(t), \\ V_p(a) = 0, \quad x(a) = x_a, \quad x(b) = x_b, \end{cases}$$

with $p = 2, \ldots, N$. Once we have a tool to transform a fractional variational problem into an integer-order one, we can go further to study more complicated problems. As a first candidate, we study fractional optimal control problems with free final time in Chapter 10. The problem is stated in the following way:

$$J[x, u, T] = \int_a^T L(t, x(t), u(t)) \, dt + \phi(T, x(T)) \longrightarrow \min,$$

subject to the control system

$$M\dot{x}(t) + N \, {}_a^C D_t^\alpha x(t) = f(t, x(t), u(t)),$$

and the initial boundary condition

$$x(a) = x_a,$$

with $(M, N) \neq (0,0)$, and x_a a fixed real number. Our goal is to generalize previous works on fractional optimal control problems by considering the end time, T, free and the dynamic control system as involving integer and fractional-order derivatives. First, we deduce necessary optimality conditions for this new problem with free end-point. Although this could be the beginning of the solution procedure, the lack of techniques to solve fractional differential equations prevent further progress. Another approach consists in using the approximation methods mentioned above, thereby converting the original problem into a classical optimal control problem that can be solved by standard computational techniques (Pooseh, Almeida and Torres, 2014).

In Chapter 11 we consider the fractional order α to be a variable $\alpha(t)$ that depends on time. This extension of the classical fractional calculus was introduced by (Samko and Ross, 1993), see also (Samko, 1995), and is nowadays recognized as a very useful tool.

Discrete-time fractional calculus of variations are considered in Chapter 12, the book ending with Chapter 13 of conclusion.

Chapter 2

The calculus of variations and optimal control

In this and the next two chapters we review the basic concepts that have an essential role in the understanding of the book. Chapter 2 starts with the notion of the calculus of variations and, without going into details, also recalls the optimal control theory, pointing out the variational approach together with main concepts, definitions, and some important results from the classical theory. A brief historical introduction to the fractional calculus is given afterwards, in Chapter 3. At the same time, we introduce the theoretical framework of the whole work, fixing notations and nomenclature. In Chapter 4, the calculus of variations and optimal control problems involving fractional operators are discussed as fractional variational problems.

2.1 The calculus of variations

Many authors trace the origins of the calculus of variations back to ancient times, the time of Dido, Queen of Carthage. Dido's problem had an intellectual nature. The question is to enclose as much land as possible within a strip of oxhide with a fixed length. As it is well known nowadays, thanks to the modern calculus of variations, the solution to Dido's problem is a circle (Kirk, 1970). Aristotle (384–322 B.C.) expresses a common belief in his *Physics* that nature follows the easiest path that requires the least amount of effort. This is the main idea behind many challenges to solve real-world problems (Berdichevsky, 2009).

2.1.1 *From light beams to the Brachistochrone problem*

Fermat believed that "nature operates by means and ways that are easiest and fastest" (Goldstine, 1980). Studying the analysis of refractions, he used

Galileo's reasoning on falling objects and claimed that in this case nature does not take the shortest path, but the one which has the least traverse time. Although the solution to this problem does not use variational methods, it had an important role in the solution of the most critical problem and the birth of the calculus of variations.

Newton also considered the problem of motion in a resisting medium, which is indeed a shape optimization problem. This problem is a well-known and well-studied example in the theory of the calculus of variations and optimal control nowadays (Gouveia, Plakhov and Torres, 2009; Plakhov and Torres, 2005; Silva and Torres, 2006). Nevertheless, the original problem, posed by Newton, was solved by only using calculus.

In 1796–1797, John Bernoulli challenged the mathematical world to solve a problem that he called the Brachistochrone problem:

> *If in a vertical plane two points A and B are given, then it is required to specify the orbit AMB of the movable point M, along which it, starting from A, and under the influence of its own weight, arrives at B in the shortest possible time. So that those who are keen of such matters will be tempted to solve this problem, it is good to know that it is not, as it may seem, purely speculative and without practical use. Rather it even appears, and this may be hard to believe, that it is very useful also for other branches of science than mechanics. In order to avoid a hasty conclusion, it should be remarked that the straight line is certainly the line of shortest distance between A and B, but it is not the one which is traveled in the shortest time. However, the curve AMB, which I shall disclose if by the end of this year nobody else has found it, is very well known among geometers (Sussmann and Willems, 2002, Sect. 2.1).*

It is not a big surprise that several responses came to this challenge. It was the time of some of the most famous mathematical minds. Solutions from John and Jakob Bernoulli were published in May 1797 together with contributions by Tschirnhaus and l'Hopital and a note from Leibniz. Newton also published a solution without a proof. Later on, other variants of this problem have been discussed by James Bernoulli.

2.1.2   *Contemporary mathematical formulation*

Having a rich history, mostly dealing with physical problems, the calculus of variations is nowadays an outstanding field with a strong mathematical formulation. Roughly speaking, the calculus of variations is the optimization of functionals.

Definition 2.1 (Functional). *A functional J is a rule of correspondence, from a vector space into its underlying scalar field, that assigns to each function x in a certain class Ω a unique number.*

The domain of a functional, Ω in Definition 2.1, is a class of functions. Suppose that x is a positive continuous function defined on the interval $[a, b]$. The area under x can be defined as a functional, i.e.,

$$J[x] = \int_a^b x(t)dt$$

is a functional that assigns to each function the area under its curve. Just like functions, for each functional, J, one can define its increment, ΔJ.

Definition 2.2 (See, e.g., (Kirk, 1970)). *Let x be a function and δx be its variation. Suppose also that the functional J is defined for x and $x + \delta x$. The increment of the functional J with respect to δx is*

$$\Delta J := J[x + \delta x] - J[x].$$

Using the notion of the increment of a functional we define its variation. The increment of J can be written as

$$\Delta J[x, \delta x] = \delta J[x, \delta x] + g(x, \delta x). \parallel \delta x \parallel,$$

where δJ is linear in δx and

$$\lim_{\parallel \delta x \parallel \to 0} g(x, \delta x) = 0.$$

In this case the functional J is said to be differentiable on x and δJ is its variation evaluated for the function x.

Now consider all functions in a class Ω for which the functional J is defined. A function x^* is a relative extremizer of J if its increment has the same sign for functions sufficiently close to x^*, i.e.,

$$\exists \epsilon > 0 \, \forall x \in \Omega \, : \, \|x - x^*\| < \epsilon \Rightarrow J[x] - J[x^*] \geq 0 \vee J[x] - J[x^*] \leq 0.$$

Note that for a relative minimizer the increment is non-negative and non-positive for a relative maximizer.

In this point, *the fundamental theorem of the calculus of variations* is used as a necessary condition to find a relative extremizer.

Theorem 2.1 (See, e.g., (Kirk, 1970)). *Let J be a differentiable functional defined in Ω. Assume also that the members x of Ω are not constrained by any boundaries. Then the variation of J, for all admissible variations of x, vanishes on an extremizer x^*.*

Many problems in the calculus of variations are included in a general problem of optimizing a definite integral of the form

$$J[x] = \int_a^b L(t, x(t), \dot{x}(t))dt, \qquad (2.1)$$

within a certain class, e.g., the class of continuously differentiable functions. In this formulation, the function L is called the Lagrangian and supposed to be twice continuously differentiable. The points a and b are called boundaries, or the initial and terminal points, respectively. The optimization is usually interpreted as a minimization or a maximization. Since these two processes are related, that is, $\max J = -\min -J$, in a theoretical context we usually discuss the minimization problem.

The problem is to find a function x with certain properties that gives a minimum value to the functional J. The function is usually assumed to pass through prescribed points, say $x(a) = x_a$ and/or $x(b) = x_b$. These are called the boundary conditions. Depending on the boundary conditions a variational problem can be classified as:

Fixed end points: the conditions at both end points are given,

$$x(a) = x_a, \qquad x(b) = x_b.$$

Free terminal point: the value of the function at the initial point is fixed and it is free at the terminal point,

$$x(a) = x_a.$$

Free initial point: the value of the function at the terminal point is fixed and it is free at the initial point,

$$x(b) = x_b.$$

Free end points: both end points are free.

Variable end points: one point and/or the other is required to be on a certain set, e.g., a prescribed curve.

Sometimes the function x is required to satisfy some constraints. Isoperimetric problems are a class of constrained variational problems for which the unknown function is needed to satisfy an integral of the form

$$\int_a^b G(t, x(t), \dot{x}(t))dt = K$$

in which $K \in \mathbb{R}$ has a fixed given value.

A variational problem can also be subjected to a dynamic constraint. In this setting, the objective is to find an extremizer x for the functional J such that an ordinary differential equation is fulfilled, i.e.,

$$\dot{x}(t) = f(t, x(t)), \quad t \in [a, b].$$

2.1.3 *Solution methods*

The aforementioned mathematical formulation allows us to derive optimality conditions for a large class of problems. The Euler–Lagrange necessary optimality condition is the key feature of the calculus of variations. This condition was introduced first by Euler in around 1744. Euler used a geometrical insight and finite differences approximations of derivatives to derive his necessary condition. Later, in 1755, Lagrange ended at the same result using analysis alone. Indeed Lagrange's work was the reason that Euler called this field the calculus of variations (Goldstine, 1980).

2.1.3.1 *Euler–Lagrange equation*

Let x be a scalar function in $C^2[a, b]$, i.e., it has a continuous first and second derivatives on the fixed interval $[a, b]$. Suppose that the Lagrangian L in (2.1) has continuous first and second partial derivatives with respect to all of its arguments. To find the extremizers of J one can use the fundamental theorem of the calculus of variations: the first variation of the functional must vanish on the extremizer. By the increment of a functional we have

$$\Delta J = J[x + \delta x] - J[x]$$
$$= \int_a^b L(t, x + \delta x, \dot{x} + \delta \dot{x}) dt - \int_a^b L(t, x, \dot{x}) dt.$$

The first integrand is expanded in a Taylor series and the terms up to the first order in δx and $\delta \dot{x}$ are kept. Finally, combining the integrals, gives the variation δJ as

$$\delta J[x, \delta x] = \int_a^b \left(\left[\frac{\partial L}{\partial x}(t, x, \dot{x}) \right] \delta x + \left[\frac{\partial L}{\partial \dot{x}}(t, x, \dot{x}) \right] \delta \dot{x} \right) dt.$$

One can now integrate by parts the term containing $\delta \dot{x}$ to obtain

$$\delta J[x, \delta x] = \left[\frac{\partial L}{\partial \dot{x}}(t, x, \dot{x}) \right] \delta x \Big|_a^b$$
$$+ \int_a^b \left(\left[\frac{\partial L}{\partial x}(t, x, \dot{x}) \right] - \frac{d}{dt} \left[\frac{\partial L}{\partial \dot{x}}(t, x, \dot{x}) \right] \right) \delta x \, dt.$$

Depending on how the boundary conditions are specified, we have different necessary conditions. In the very simple form when the problem is in the fixed end-points form, $\delta x(a) = \delta x(b) = 0$, the terms outside the integral vanish. For the first variation to be vanished one has

$$\int_a^b \left(\frac{\partial L}{\partial x}(t, x, \dot{x}) - \frac{d}{dt} \left[\frac{\partial L}{\partial \dot{x}}(t, x, \dot{x}) \right] \right) \delta x \, dt = 0.$$

According to the *fundamental lemma of the calculus of variations* (see, e.g., (van Brunt, 2004)), if a function h is continuous and

$$\int_a^b h(t)\eta(t)dt = 0,$$

for every function η that is continuous in the interval $[a, b]$, then h must be zero everywhere in the interval $[a, b]$. Therefore, the Euler–Lagrange necessary optimality condition, that is an ordinary differential equation, reads to

$$\frac{\partial L}{\partial x}(t, x, \dot{x}) - \frac{d}{dt}\left[\frac{\partial L}{\partial \dot{x}}(t, x, \dot{x})\right] = 0,$$

when the boundary conditions are given at both end-points. For free endpoint problems the so-called transversality conditions are added to the Euler–Lagrange equation (see, e.g., (Malinowska and Torres, 2010a)).

Definition 2.3. *Solutions to the Euler–Lagrange equation are called extremals for J defined by (2.1).*

The necessary condition for optimality can also be derived using the classical method of perturbing the extremal and using the Gateaux derivative. The Gateaux differential or Gateaux derivative is a generalization of the concept of directional derivative:

$$dF[x; \eta] = \lim_{\epsilon \to 0}\frac{F[x + \epsilon\eta] - F[x]}{\epsilon} = \frac{d}{d\epsilon}F[x + \epsilon\eta]\Big|_{\epsilon=0}.$$

Let $x^* \in C^2[a, b]$ be the extremal and $\eta \in C^2[a, b]$ be such that $\eta(a) = \eta(b) = 0$. Then for sufficiently small values of ϵ, form the family of curves $x^* + \epsilon\eta$. All of these curves reside in a neighborhood of x^* and are admissible functions, i.e., they are in the class Ω and satisfy the boundary conditions. We now construct the function

$$j(\epsilon) = \int_a^b L(t, x^*(t) + \epsilon\eta(t), \dot{x}^*(t) + \epsilon\dot{\eta}(t))dt, \quad -\delta < \epsilon < \delta. \qquad (2.2)$$

Due to the construction of the function $j(\epsilon)$, the extremum is achieved for $\epsilon = 0$. Therefore, it is necessary that the first derivative of $j(\epsilon)$ vanishes for $\epsilon = 0$, i.e.,

$$\frac{dj(\epsilon)}{d\epsilon}\Big|_{\epsilon=0} = 0.$$

Differentiating (2.2) with respect to ϵ, we get

$$\frac{dj(\epsilon)}{d\epsilon} = \int_a^b \left(\left[\frac{\partial L}{\partial x}(t, x^* + \epsilon\eta, \dot{x}^* + \epsilon\dot{\eta}) \right] \eta + \left[\frac{\partial L}{\partial \dot{x}}(t, x^* + \epsilon\eta, \dot{x}^* + \epsilon\dot{\eta}) \right] \dot{\eta} \right) dt.$$

Setting $\epsilon = 0$, we arrive at the formula

$$\int_a^b \left(\left[\frac{\partial L}{\partial x}(t, x^*, \dot{x}^*) \right] \eta + \left[\frac{\partial L}{\partial \dot{x}}(t, x^*, \dot{x}^*) \right] \dot{\eta} \right) dt,$$

which gives the Euler–Lagrange condition after making an integration by parts and applying the fundamental lemma.

The solution to the Euler–Lagrange equation, if exists, is an extremal for the variational problem. Except for simple problems, it is very difficult to solve such differential equations in a closed form. Therefore, numerical methods are employed for most practical purposes.

2.1.3.2 *Numerical methods*

A variational problem can be solved numerically in two different ways: by indirect or direct methods. Constructing the Euler–Lagrange equation and solving the resulting differential equation is known to be the indirect method.

There are two main classes of direct methods. On one hand, we specify a discretization scheme by choosing a set of mesh points on the horizon of interest, say $a = t_0, t_1, \ldots, t_n = b$ for $[a, b]$. Then we use some approximations for derivatives in terms of the unknown function values at t_i and using an appropriate quadrature, the problem is transformed to a finite-dimensional optimization. This method is known as Euler's method in the literature. Regarding Fig. 2.1, the solid line is the function that we are looking for, nevertheless, the method gives the polygonal dashed line as an approximate solution.

On the other hand, there is the Ritz method, that has an extension to functionals of several independent variables which is called Kantorovich's method. We assume that the admissible functions can be expanded in some kind of series, e.g. power or Fourier's series, of the form

$$x(t) = \sum_{k=0}^{\infty} a_k \phi_k(t).$$

Using a finite number of terms in the sum as an approximation, and some sort of quadrature again, the original problem can be transformed to an equivalent optimization problem for a_k, $k = 0, 1, \ldots, n$.

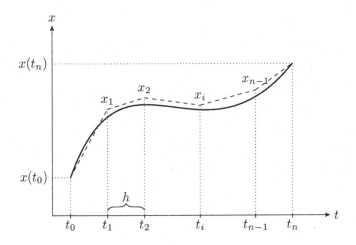

Fig. 2.1   Euler's finite differences method.

2.2   Optimal control theory

Optimal control theory is a well-studied subject. Many papers and text-books present the field very well, see (Bryson and Ho, 1975; Kirk, 1970; Pontryagin *et al.*, 1962). Nevertheless, we introduce some basic concepts without going into details. Our main purpose is to review the variational approach to optimal control theory and clarify its connection to the calculus of variations. This provides a background for our later investigations on fractional variational problems. The formulation is presented for vector functions, $\mathbf{x} = (x_1, \ldots, x_n)$, to emphasize the possibility of such functions. This is also valid, and easy to adapt, for the calculus of variations.

2.2.1   *Mathematical formulation*

Mathematically speaking, the notion of control is highly connected to dynamical systems. A dynamical system is usually formulated using a system of ordinary or partial differential equations. In this book, dealing only with ordinary derivatives, we consider the dynamics as

$$\begin{cases} \dot{\mathbf{x}} = f(t, \mathbf{x}), \\ \mathbf{x}(t_0) = \mathbf{x}_0, \end{cases}$$

where $\mathbf{x} = (x_1, \ldots, x_n)$, the state of the system, is a vector function, $t_0 \in \mathbb{R}$, $\mathbf{x}_0 \in \mathbb{R}^n$ and $f : \mathbb{R}^{n+1} \to \mathbb{R}^n$ are given.

In order to affect the behavior of a system, e.g., a real-life physical system used in technology, one can introduce control parameters to the system. A controlled system also can be described by a system of ordinary differential equations,

$$\begin{cases} \dot{\mathbf{x}} = f(t, \mathbf{x}, \mathbf{u}), \\ \mathbf{x}(t_0) = \mathbf{x}_0, \end{cases}$$

in which $\mathbf{u} \in \Omega \subseteq \mathbb{R}^m$ is the control parameter or variable. The control parameters can also be time-varying, i.e., $\mathbf{u} = \mathbf{u}(t)$. In this case $f : \mathbb{R}^{n+m+1} \to \mathbb{R}^n$ is supposed to be continuous with respect to all of its arguments and continuously differentiable with respect to $\mathbf{x} = (x_1, \ldots, x_n)$.

In an optimal control problem, the main objective is to determine the control parameters in a way that certain optimality criteria are fulfilled. In this book we consider problems in which a functional of the form

$$J[\mathbf{x}, \mathbf{u}] = \int_a^b L(t, \mathbf{x}(t), \mathbf{u}(t)) dt$$

should be optimized. Therefore, a typical optimal control problem is formulated as

$$J[\mathbf{x}, \mathbf{u}] = \int_a^b L(t, \mathbf{x}(t), \mathbf{u}(t)) dt \longrightarrow \min$$

$$s.t. \quad \begin{cases} \dot{\mathbf{x}}(t) = f(t, \mathbf{x}(t), \mathbf{u}(t)) \\ \mathbf{x}(t_0) = \mathbf{x}_0, \end{cases}$$

where the state $\mathbf{x}$ and the control $\mathbf{u}$ are assumed to be unbounded. This formulation can also be considered as a framework for both optimal control and the calculus of variations. Let $\dot{x}(t) = u(t)$. Then the optimization of (2.1) becomes

$$J[x] = \int_a^b L(t, x(t), u(t)) dt \longrightarrow \min$$

$$s.t. \quad \begin{cases} \dot{x}(t) = u(t) \\ x(t_0) = x_0, \end{cases}$$

that is an optimal control problem. On one hand, we can apply aforementioned direct methods. On the other hand, indirect methods consist in using Lagrange multipliers in a variational approach to obtain the Euler–Lagrange equations. The dynamics is considered as a constraint for a variational problem and is added into the functional. The so-called augmented functional is then achieved, that is, the functional

$$J_a[\mathbf{x}, \mathbf{u}] = \int_a^b \left[L(t, \mathbf{x}(t), \mathbf{u}(t)) - \boldsymbol{\lambda}(t)^T (\dot{\mathbf{x}}(t) - f(t, \mathbf{x}(t), \mathbf{u}(t))) \right] dt$$

is treated subject to the boundary conditions.

2.2.2   Necessary optimality conditions

Although the Euler–Lagrange equations are derived by usual ways, e.g., Section 2.1.3, it is common and useful to define the Hamiltonian function by

$$H(t, \mathbf{x}, \mathbf{u}, \boldsymbol{\lambda}) = L(t, \mathbf{x}, \mathbf{u}) + \boldsymbol{\lambda}^T [f(t, \mathbf{x}, \mathbf{u})].$$

Then the necessary optimality conditions read as

$$\begin{cases} \dot{\mathbf{x}}(t) = \dfrac{\partial H}{\partial \boldsymbol{\lambda}}(t, \mathbf{x}(t), \mathbf{u}(t), \boldsymbol{\lambda}(t)), \\[2mm] \dot{\boldsymbol{\lambda}}(t) = -\dfrac{\partial H}{\partial \mathbf{x}}(t, \mathbf{x}(t), \mathbf{u}(t), \boldsymbol{\lambda}(t)), \\[2mm] 0 = \dfrac{\partial H}{\partial \mathbf{u}}(t, \mathbf{x}(t), \mathbf{u}(t), \boldsymbol{\lambda}(t)). \end{cases}$$

It is possible to consider a function $\phi(b, \mathbf{x}(b))$ in the objective functional, which makes the cost functional dependent on the time and state variables at the terminal point. This can be treated easily by some more calculations. Also one can discuss different end-point conditions in the same way as we did for the calculus of variations.

2.2.3   Pontryagin's minimum principle

Roughly speaking, unbounded control is an essential assumption to use variational methods freely and to obtain the resulting necessary optimality conditions. In contrast, if there is a bound on control, then δu can no longer vary freely. Therefore, the fact that δJ must vanish on a extremal is of no use. Nevertheless, special variations can be defined and used to prove that for u^* to be an extremal, it is necessary that

$$H(t, x^*, u^* + \delta u, \lambda^*) \geq H(t, x^*, u^*, \lambda^*),$$

for all admissible δu (Pontryagin *et al.*, 1962). That is, an optimal control u^* is a global minimizer of the Hamiltonian for a control system. This condition is known as Pontryagin's minimum principle. It is worthwhile to note that the condition that the partial derivative of the Hamiltonian with respect to control u must vanish on an optimal control is a necessary condition for the minimum principle:

$$\frac{\partial H}{\partial u}(t, x^*, u^*, \lambda^*) = 0.$$

Chapter 3

Fractional calculus

In the early ages of modern differential calculus, right after the introduction of $\frac{d}{dt}$ for the first derivative, in a letter dated 1695, l'Hopital asked Leibniz the meaning of $\frac{d^{\frac{1}{2}}}{dt^{\frac{1}{2}}}$, the derivative of order $\frac{1}{2}$ (Miller and Ross, 1993). The appearance of $\frac{1}{2}$ as a fraction gave the name fractional calculus to the study of derivatives, and integrals, of any order, real or complex.

There are several different approaches and definitions in fractional calculus for derivatives and integrals of arbitrary order. Here we give a historical progress of the theory of fractional calculus that includes all we need throughout the book. We mostly follow the notation used in the books (Kilbas, Srivastava and Trujillo, 2006; Samko, Kilbas and Marichev, 1993). Before getting into the details of the theory, we briefly outline the definitions of some special functions that are used in the definitions of fractional derivatives and integrals, or appear in some manipulation, e.g., solving fractional differential and integral equations.

3.1 Special functions

Although there are many special functions that appear in fractional calculus, in this book, only a few of them are encountered. The following definitions are introduced together with some properties.

Definition 3.1 (Gamma function). *The Euler integral of the second kind*

$$\Gamma(z) = \int_0^\infty t^{z-1} e^{-t} dt, \qquad \Re(z) > 0,$$

is called the Gamma function.

The Gamma function has an important property, $\Gamma(z + 1) = z\Gamma(z)$ and hence $\Gamma(z) = z!$ for $z \in \mathbb{N}$, that allows us to extend the notion of factorial to real numbers. For further properties of this special function we refer the reader to (Andrews, Askey and Roy, 1999).

Remark 3.1. There are several alternative definitions for the Gamma function. The following infinite product definition is valid for all complex numbers z, except the non-positive integers:

$$\Gamma(z) = \frac{1}{z} \prod_{n=1}^{\infty} \frac{\left(1 + \frac{1}{n}\right)^z}{1 + \frac{z}{n}}. \tag{3.1}$$

Definition 3.2 (Mittag–Leffler function). *Let* $\alpha > 0$. *The function* E_α *defined by*

$$E_\alpha(z) = \sum_{j=0}^{\infty} \frac{z^j}{\Gamma(\alpha j + 1)},$$

whenever the series converges, is called the one parameter Mittag–Leffler function. The two-parameter Mittag–Leffler function with parameters α, $\beta > 0$ *is defined by*

$$E_{\alpha,\beta}(z) = \sum_{j=0}^{\infty} \frac{z^j}{\Gamma(\alpha j + \beta)}.$$

The Mittag–Leffler function is a generalization of exponential series and coincides with the series expansion of e^z for $\alpha = 1$ (and $\beta = 1$).

3.2   A historical review

Attempting to answer the question of l'Hopital, Leibniz tried to explain the possibility of the derivative of order $\frac{1}{2}$. He also stated that "this is an apparent paradox from which, one day, useful consequences will be drawn" (Kilbas, Srivastava and Trujillo, 2006, p. vii). During the next century the question was raised again by Euler (1738), expressing an interest to the calculation of fractional-order derivatives.

The nineteenth century has witnessed much effort in the field. In 1812, Laplace discussed non-integer derivatives of some functions that are representable by integrals. Later, in 1819, Lacriox generalized $\frac{d^n}{dt^n}t^n$ to $\frac{d^{\frac{1}{2}}}{dt^{\frac{1}{2}}}t$.

The first challenge of making a definition for arbitrary order derivatives comes from Fourier in 1822, with

$$\frac{d^\alpha}{dt^\alpha}x(t) = \frac{1}{2\pi}\int_{-\infty}^{\infty} x(\tau)d\tau \int_{-\infty}^{\infty} p^\alpha \cos\left[p(t-\tau) + \frac{1}{2}\alpha\pi\right]dp.$$

He derived this definition from the integral representation of a function x. The next important step was taken by Abel in 1823. Solving the Tautochrone problem, he worked with integral equations of the form

$$\int_0^t (t-\tau)^{-\alpha}x(\tau)d\tau = k.$$

Apart from a multiplicative factor, the left-hand side of this equation resembles the modern definitions of fractional derivatives. Almost ten years later the first definitions of fractional operators appeared in the works of Liouville (1832), and this has been contributed to by many other mathematicians like Peacock and Kelland (1839), and Gregory (1841). Finally, starting from 1847, Riemann dedicated some work to fractional integrals that led to the introduction of Riemann–Liouville fractional derivatives and integrals by Sonin in 1869.

Definition 3.3 (Riemann–Liouville fractional integral). *Let x be an integrable function in $[a,b]$ and $\alpha > 0$.*

- *The left Riemann–Liouville fractional integral of order α is given by*

$$_aI_t^\alpha x(t) = \frac{1}{\Gamma(\alpha)}\int_a^t (t-\tau)^{\alpha-1}x(\tau)d\tau, \quad t > a.$$

- *The right Riemann–Liouville fractional integral of order α is given by*

$$_tI_b^\alpha x(t) = \frac{1}{\Gamma(\alpha)}\int_t^b (\tau-t)^{\alpha-1}x(\tau)d\tau, \quad t < b.$$

Definition 3.4 (Riemann–Liouville fractional derivative). *Let x be an absolutely continuous function in $[a,b]$, $\alpha > 0$, and $n = [\alpha] + 1$.*

- *The left Riemann–Liouville fractional derivative of order α is given by*

$$_aD_t^\alpha x(t) = \frac{1}{\Gamma(n-\alpha)}\left(\frac{d}{dt}\right)^n \int_a^t (t-\tau)^{n-1-\alpha}x(\tau)d\tau, \quad t > a.$$

- *The right Riemann–Liouville fractional derivative of order α is given by*

$$_tD_b^\alpha x(t) = \frac{1}{\Gamma(n-\alpha)} \left(-\frac{d}{dt}\right)^n \int_t^b (\tau - t)^{n-1-\alpha} x(\tau)d\tau, \quad t < b.$$

These definitions are motivated by Cauchy's n-fold integral formula:

$$I^n x(t) = \int_0^t \int_0^{t_{n-1}} \cdots \int_0^{t_1} x(t_0)dt_0 dt_1 \ldots dt_{n-1}$$

$$= \frac{1}{(n-1)!} \int_0^t (t-\tau)^{n-1} x(\tau)d\tau. \tag{3.2}$$

Substituting n by α in (3.2) and using the Gamma function, $\Gamma(n) = (n-1)!$, leads to Definition 3.3:

$$I^\alpha x(t) = \frac{1}{\Gamma(\alpha)} \int_0^t (t-\tau)^{\alpha-1} x(\tau)d\tau.$$

For the derivative, one has $D^\alpha x(t) = D^n I^{n-\alpha} x(t)$.

The next important definition is a generalization of the definition of higher-order derivatives and appeared in the works of Grünwald (1867) and Letnikov (1868).

In classical theory, given a derivative of certain order $x^{(n)}$, there is a finite difference approximation of the form

$$x^{(n)}(t) = \lim_{h\to 0^+} \frac{1}{h^n} \sum_{k=0}^n (-1)^k \binom{n}{k} x(t - kh), \tag{3.3}$$

where $\binom{n}{k}$ is the binomial coefficient, that is,

$$\binom{n}{k} = \frac{n(n-1)(n-2)\cdots(n-k+1)}{k!}, \quad n, k \in \mathbb{N}.$$

In what follows $\binom{\alpha}{k}$ is the generalization of binomial coefficients to real numbers, that is,

$$\binom{\alpha}{k} = \frac{\Gamma(\alpha+1)}{\Gamma(k+1)\Gamma(\alpha-k+1)},$$

where k and α can be any integer, real or complex, except that $\alpha \notin \{-1, -2, -3, \ldots\}$. The Grünwald–Letnikov definition of fractional derivative is a generalization of formula (3.3) to derivatives of arbitrary order.

Definition 3.5 (Grünwald–Letnikov derivative). *Let $\alpha > 0$.*

- *The left Grünwald–Letnikov fractional derivative is defined as*

$$
{}^{GL}_{a}D^{\alpha}_{t}x(t) = \lim_{h\to 0^{+}} \frac{1}{h^{\alpha}} \sum_{k=0}^{\infty} (-1)^{k} \binom{\alpha}{k} x(t - kh). \qquad (3.4)
$$

- *The right Grünwald–Letnikov derivative is*

$$
{}^{GL}_{t}D^{\alpha}_{b}x(t) = \lim_{h\to 0^{+}} \frac{1}{h^{\alpha}} \sum_{k=0}^{\infty} (-1)^{k} \binom{\alpha}{k} x(t + kh). \qquad (3.5)
$$

The series in (3.4) and (3.5), the Grünwald–Letnikov definitions, converge absolutely and uniformly if x is bounded. The infinite sums, backward differences for left and forward differences for right derivatives in the Grünwald–Letnikov definitions of fractional derivatives, reveal that the arbitrary order derivative of a function at a time t depends on all values of that function in $(-\infty, t]$ and $[t, \infty)$ for left and right derivatives, respectively. This is due to the non-local property of fractional derivatives.

Remark 3.2. Equations (3.4) and (3.5) need to be consistent in closed time intervals and we need the values of $x(t)$ outside the interval $[a, b]$. To overcome this difficulty, we can take

$$
x^{*}(t) = \begin{cases} x(t) & t \in [a, b], \\ 0 & t \notin [a, b]. \end{cases}
$$

Then we assume ${}^{GL}_{a}D^{\alpha}_{t}x(t) = {}^{GL}_{a}D^{\alpha}_{t}x^{*}(t)$ and ${}^{GL}_{t}D^{\alpha}_{b}x(t) = {}^{GL}_{t}D^{\alpha}_{b}x^{*}(t)$ for $t \in [a, b]$.

These definitions coincide with the definitions of Riemann–Liouville derivatives.

Proposition 3.1 (See (Podlubny, 1999)). *Let $0 < \alpha < n$, $n \in \mathbb{N}$ and $x \in C^{n-1}[a, b]$. Suppose that $x^{(n)}$ is integrable on $[a, b]$. Then for every α the Riemann–Liouville derivative exists and coincides with the Grünwald–Letnikov derivative:*

$$
{}_{a}D^{\alpha}_{t}x(t) = \sum_{i=0}^{n-1} \frac{x^{(i)}(a)(t-a)^{i-\alpha}}{\Gamma(1+i-\alpha)} + \frac{1}{\Gamma(n-\alpha)} \int_{a}^{t} (t-\tau)^{n-1-\alpha} x^{(n)}(\tau) d\tau
$$

$$
= {}^{GL}_{a}D^{\alpha}_{t}x(t).
$$

Another type of fractional operators, that is investigated in this book, is the Hadamard type operators introduced in 1892 (Hadamard, 1892).

Definition 3.6 (Hadamard fractional integral). *Let a, b be two real numbers with $0 < a < b$ and $x : [a, b] \to \mathbb{R}$.*

- *The left Hadamard fractional integral of order $\alpha > 0$ is defined by*

$$_a\mathcal{I}_t^\alpha x(t) = \frac{1}{\Gamma(\alpha)} \int_a^t \left(\ln \frac{t}{\tau} \right)^{\alpha-1} \frac{x(\tau)}{\tau} d\tau, \quad t > a.$$

- *The right Hadamard fractional integral of order $\alpha > 0$ is defined by*

$$_t\mathcal{I}_b^\alpha x(t) = \frac{1}{\Gamma(\alpha)} \int_t^b \left(\ln \frac{\tau}{t} \right)^{\alpha-1} \frac{x(\tau)}{\tau} d\tau, \quad t < b.$$

When $\alpha = m$ is an integer, these fractional integrals are m-fold integrals:

$$_a\mathcal{I}_t^m x(t) = \int_a^t \frac{d\tau_1}{\tau_1} \int_a^{\tau_1} \frac{d\tau_2}{\tau_2} \cdots \int_a^{\tau_{m-1}} \frac{x(\tau_m)}{\tau_m} d\tau_m$$

and

$$_t\mathcal{I}_b^m x(t) = \int_t^b \frac{d\tau_1}{\tau_1} \int_{\tau_1}^b \frac{d\tau_2}{\tau_2} \cdots \int_{\tau_{m-1}}^b \frac{x(\tau_m)}{\tau_m} d\tau_m.$$

For fractional derivatives, we also consider the left and right derivatives.

Definition 3.7 (Hadamard fractional derivative). *Let $\alpha > 0$ be a real number and $n = [\alpha] + 1$.*

- *The left Hadamard fractional derivative of order α is defined by*

$$_a\mathcal{D}_t^\alpha x(t) = \left(t \frac{d}{dt} \right)^n \frac{1}{\Gamma(n-\alpha)} \int_a^t \left(\ln \frac{t}{\tau} \right)^{n-\alpha-1} \frac{x(\tau)}{\tau} d\tau, \quad t > a.$$

- *The right Hadamard fractional derivative of order α is defined by*

$$_t\mathcal{D}_b^\alpha x(t) = \left(-t \frac{d}{dt} \right)^n \frac{1}{\Gamma(n-\alpha)} \int_t^b \left(\ln \frac{\tau}{t} \right)^{n-\alpha-1} \frac{x(\tau)}{\tau} d\tau, \quad t < b.$$

When $\alpha = m$ is an integer, we have

$$_a\mathcal{D}_t^m x(t) = \left(t \frac{d}{dt} \right)^m x(t) \text{ and } {_t\mathcal{D}_b^m} x(t) = \left(-t \frac{d}{dt} \right)^m x(t).$$

Finally, we recall another definition, the Caputo derivatives, that are believed to be more applicable in practical fields such as engineering and physics. In spite of the success of the Riemann–Liouville approach in theory, some difficulties arise in practice where initial conditions need to be treated, for instance, in fractional differential equations. Such conditions for Riemann–Liouville case have no clear physical interpretations (Podlubny, 1999). The following definition was proposed by Caputo in 1967. Caputo's fractional derivatives are, however, related to Riemann–Liouville definitions.

Definition 3.8 (Caputo's fractional derivatives). *Let $\alpha > 0$ be a real number, $n = [\alpha] + 1$ and $x \in AC^n[a, b]$.*

- *The left Caputo fractional derivative of order α is given by*

$$_a^C D_t^\alpha x(t) = \frac{1}{\Gamma(n - \alpha)} \int_a^t (t - \tau)^{n-1-\alpha} x^{(n)}(\tau)d\tau, \quad t > a.$$

- *The right Caputo fractional derivative of order α is given by*

$$_t^C D_b^\alpha x(t) = \frac{(-1)^n}{\Gamma(n - \alpha)} \int_t^b (\tau - t)^{n-1-\alpha} x^{(n)}(\tau)d\tau, \quad t < b.$$

These fractional integrals and derivatives define a rich calculus. For details see the books (Kilbas, Srivastava and Trujillo, 2006; Miller and Ross, 1993; Samko, Kilbas and Marichev, 1993). Here we just recall some useful properties for our purposes.

3.3 The relation between the Riemann–Liouville and Caputo derivatives

For $\alpha > 0$ and $n = [\alpha] + 1$, the Riemann–Liouville and Caputo derivatives are related by the following formulas:

$$_a D_t^\alpha x(t) = {_a^C D_t^\alpha} x(t) + \sum_{k=0}^{n-1} \frac{x^{(k)}(a)}{\Gamma(k + 1 - \alpha)} (t - a)^{k-\alpha} \tag{3.6}$$

and

$$_t D_b^\alpha x(t) = {_t^C D_b^\alpha} x(t) + \sum_{k=0}^{n-1} \frac{x^{(k)}(b)}{\Gamma(k + 1 - \alpha)} (b - t)^{k-\alpha}. \tag{3.7}$$

Actually, properties (3.6) and (3.7) coincide with the original definition of Caputo derivative, given by M. Caputo in his 1967 paper (Caputo, 1967).

In some cases the two derivatives coincide,

$$_a D_t^\alpha x = {_a^C D_t^\alpha} x, \quad \text{when } x^{(k)}(a) = 0, \quad k = 0, \ldots, n - 1,$$

$$_t D_b^\beta x = {_t^C D_b^\beta} x, \quad \text{when } x^{(k)}(b) = 0, \quad k = 0, \ldots, n - 1.$$

3.4 Integration by parts

Formulas of integration by parts have an important role in the proof of Euler–Lagrange necessary optimality conditions.

Lemma 3.1 (cf. (Kilbas, Srivastava and Trujillo, 2006)).
Let $\alpha > 0$, $p, q \geq 1$ and $\frac{1}{p} + \frac{1}{q} \leq 1 + \alpha$ ($p \neq 1$ and $q \neq 1$ in the case where $\frac{1}{p} + \frac{1}{q} = 1 + \alpha$).
(i) If $\varphi \in L_p(a, b)$ and $\psi \in L_q(a, b)$, then

$$\int_a^b \varphi(t) {}_a I_t^\alpha \psi(t) dt = \int_a^b \psi(t) {}_t I_b^\alpha \varphi(t) dt.$$

(ii) If $g \in {}_t I_b^\alpha(L_p)$ and $f \in {}_a I_t^\alpha(L_q)$, then

$$\int_a^b g(t) {}_a D_t^\alpha f(t) dt = \int_a^b f(t) {}_t D_b^\alpha g(t) dt,$$

where the space of functions ${}_t I_b^\alpha(L_p)$ and ${}_a I_t^\alpha(L_q)$ are defined for $\alpha > 0$ and $1 \leq p \leq \infty$ by

$$ {}_a I_t^\alpha(L_p) := \{f : f = {}_a I_t^\alpha \varphi, \ \varphi \in L_p(a, b)\} $$

and

$$ {}_t I_b^\alpha(L_p) := \{f : f = {}_t I_b^\alpha \varphi, \ \varphi \in L_p(a, b)\}. $$

For Caputo fractional derivatives,

$$\int_a^b g(t) {\cdot}_a^C D_t^\alpha f(t) dt = \int_a^b f(t) {\cdot}_t D_b^\alpha g(t) dt + \sum_{j=0}^{n-1} \left[{}_t D_b^{\alpha+j-n} g(t) \cdot f^{(n-1-j)}(t) \right]_a^b$$

(see, e.g., (Agrawal, 2007a, Eq. (16))). In particular, for $\alpha \in (0, 1)$ one has

$$\int_a^b g(t) {\cdot}_a^C D_t^\alpha f(t) dt = \int_a^b f(t) \cdot {}_t D_b^\alpha g(t) dt + \left[{}_t I_b^{1-\alpha} g(t) \cdot f(t) \right]_a^b. \qquad (3.8)$$

When $\alpha \to 1$, ${}_a^C D_t^\alpha = \frac{d}{dt}$, ${}_t D_b^\alpha = -\frac{d}{dt}$, ${}_t I_b^{1-\alpha}$ is the identity operator, and (3.8) gives the classical formula of integration by parts.

Chapter 4

Fractional variational problems

A fractional problem of the calculus of variations and optimal control consists in the study of an optimization problem, in which the objective functional or constraints depend on derivatives and/or integrals of arbitrary, real or complex, orders. This is a generalization of the classical theory, where derivatives and integrals can only appear in integer orders. The aim of this chapter is to present necessary conditions that every extremizer of a functional must fulfill. Usually such equations involve fractional operators, and are called (fractional) Euler–Lagrange equations. Solving such fractional equations we obtain a list of possible solutions for the problem. In the literature we can find several distinct types of functionals, involving distinct fractional operators, and thus different Euler–Lagrange equations are deduced (Malinowska and Torres, 2012). This chapter is essentially based on the original results published in (Almeida, 2012, 2013).

4.1 Fractional calculus of variations and optimal control

Many generalizations of the classical calculus of variations and optimal control have been made, in order to extend the theory to the field of fractional variational and fractional optimal control. A simple fractional variational problem consists in finding a function x that minimizes the functional

$$J[x] = \int_a^b L(t, x(t), {_aD_t^\alpha}x(t))dt, \qquad (4.1)$$

where ${_aD_t^\alpha}$ is the left Riemann–Liouville fractional derivative. Typically, some boundary conditions are prescribed as $x(a) = x_a$ and/or $x(b) = x_b$. Classical techniques have been adopted to solve such problems. The Euler–Lagrange equation for a Lagrangian of the form $L(t, x(t), {_aD_t^\alpha}x(t), {_tD_b^\alpha}x(t))$

has been derived in (Agrawal, 2002). Many variants of necessary conditions of optimality have been studied. A generalization of the problem to include fractional integrals, i.e., $L = L(t, {}_aI_t^{1-\alpha}x(t), {}_aD_t^\alpha x(t))$, the transversality conditions of fractional variational problems and many other aspects can be found in the literature of recent years. See (Almeida and Torres, 2009a, 2010; Atanacković, Konjik and Pilipović, 2008; Riewe, 1996, 1997) and references therein. Furthermore, it has been shown that a variational problem with fractional derivatives can be reduced to a classical problem using an approximation of the Riemann–Liouville fractional derivatives in terms of a finite sum, where only derivatives of integer order are present (Atanacković, Konjik and Pilipović, 2008).

On the other hand, fractional optimal control problems usually appear in the form of

$$J[x] = \int_a^b L(t, x(t), u(t))dt \longrightarrow \min$$

subject to

$$\begin{cases} {}_aD_t^\alpha x(t) = f(t, x(t), u(t)) \\ x(a) = x_a, \ x(b) = x_b, \end{cases}$$

where an optimal control u together with an optimal trajectory x are required to follow a fractional dynamics and, at the same time, optimize an objective functional. Again, classical techniques are generalized to derive necessary optimality conditions. Euler–Lagrange equations have been introduced, e.g., in (Agrawal, 2004). A Hamiltonian formalism for fractional optimal control problems can be found in (Baleanu, Defterli and Agrawal, 2009) that exactly follows the same procedure of the regular optimal control theory, i.e., those with only integer-order derivatives.

4.2   A general formulation

The appearance of fractional terms of different types, derivatives and integrals, and the fact that there are several definitions for such operators, makes it difficult to present a typical problem to represent all possibilities. Nevertheless, one can consider the optimization of functionals of the form

$$J[\mathbf{x}] = \int_a^b L(t, \mathbf{x}(t), D^\alpha \mathbf{x}(t))dt \tag{4.2}$$

that depends on the fractional derivative, D^α, in which $\mathbf{x} = (x_1, \ldots, x_n)$ is a vector function, $\boldsymbol{\alpha} = (\alpha_1, \ldots, \alpha_n)$ and α_i, $i = 1, \ldots, n$ are arbitrary real numbers. The problem can be with or without boundary conditions. Many settings of fractional variational and optimal control problems can be transformed into the optimization of (4.2). Constraints that usually appear in the calculus of variations and are always present in optimal control problems can be included in the functional using Lagrange multipliers. More precisely, in presence of dynamic constraints as fractional differential equations, we assume that it is possible to transform such equations to a vector fractional differential equation of the form

$$D^{\alpha}\mathbf{x}(t) = f(t, \mathbf{x}(t)).$$

In this stage, we introduce a new variable $\boldsymbol{\lambda} = (\lambda_1, \lambda_2, \ldots, \lambda_n)$ and consider the optimization of

$$J[\mathbf{x}] = \int_a^b \left[L(t, \mathbf{x}(t), D^{\alpha}\mathbf{x}(t)) - \boldsymbol{\lambda}(t)D^{\alpha}\mathbf{x}(t) + \boldsymbol{\lambda}(t)f(t, \mathbf{x}(t)) \right] dt.$$

When the problem depends on fractional integrals, I^{α}, a new variable can be defined as $z(t) = I^{\alpha}x(t)$. Recall that $D^{\alpha}I^{\alpha}x = x$, see (Kilbas, Srivastava and Trujillo, 2006). The equation

$$D^{\alpha}z(t) = D^{\alpha}I^{\alpha}x(t) = x(t),$$

can be regarded as an extra constraint to be added to the original problem. However, problems containing fractional integrals can be treated directly to avoid the complexity of adding an extra variable to the original problem. Interested readers are addressed to (Almeida and Torres, 2010; Pooseh, Almeida and Torres, 2012a).

Throughout this book, by a fractional variational problem, we mainly consider the following one variable problem with given boundary conditions:

$$J[x] = \int_a^b L(t, x(t), D^{\alpha}x(t))dt \longrightarrow \min$$

subject to

$$\begin{cases} x(a) = x_a \\ x(b) = x_b. \end{cases}$$

In this setting, D^{α} can be replaced by any fractional operator that is available in literature, say, Riemann–Liouville, Caputo, Grünwald–Letnikov, Hadamard and so forth. The inclusion of a constraint is done by Lagrange multipliers. The transition from this problem to the general one, Eq. (4.2), is straightforward and is not discussed here.

4.3   Fractional Euler–Lagrange equations

Many generalizations to the classical calculus of variations have been made in recent years, in order to extend the theory to the field of fractional variational problems. As an example, consider the following minimization problem:

$$J[x] = \int_a^b L(t, x(t), {}_aD_t^\alpha x(t))dt \longrightarrow \min$$
$$s.t. \ x(a) = x_a, \ x(b) = x_b,$$

where $x \in AC[a, b]$ and L is a smooth function of t.

Using the classical methods we can obtain the following theorem as the necessary optimality condition for the fractional calculus of variations. We refer the reader to Section 2.2.1 of (Malinowska and Torres, 2012).

Theorem 4.1 (cf. (Agrawal, 2002)). *Let J be a functional of the form*

$$J[x] = \int_a^b L(t, x(t), {}_aD_t^\alpha x(t))dt, \tag{4.3}$$

defined on the set of functions x which have continuous left and right Riemann–Liouville derivatives of order α in $[a, b]$, and satisfy the boundary conditions $x(a) = x_a$ and $x(b) = x_b$. A necessary condition for J to have an extremum for a function x is that x satisfy the following Euler–Lagrange equation:

$$\frac{\partial L}{\partial x} + {}_tD_b^\alpha \left(\frac{\partial L}{\partial_aD_t^\alpha x} \right) = 0.$$

Proof.   Assume that x^* is the optimal solution. Let $\epsilon \in \mathbb{R}$ and define a family of functions

$$x(t) = x^*(t) + \epsilon\eta(t)$$

which satisfy the boundary conditions. So one should have $\eta(a) = \eta(b) = 0$. Since ${}_aD_t^\alpha$ is a linear operator, it follows that

$${}_aD_t^\alpha x(t) = {}_aD_t^\alpha x^*(t) + \epsilon \, {}_aD_t^\alpha \eta(t).$$

Substituting in (4.3) we find that for each η

$$j(\epsilon) = \int_a^b L(t, x^*(t) + \epsilon\eta(t), {}_aD_t^\alpha x^*(t) + \epsilon \, {}_aD_t^\alpha \eta(t))dt$$

is a function of ϵ only. Note that $j(\epsilon)$ has an extremum at $\epsilon = 0$. Differentiating with respect to ϵ (the Gateaux derivative) we conclude that

$$\frac{dj}{d\epsilon}\Big|_{\epsilon=0} = \int_a^b \left(\frac{\partial L}{\partial x}\eta + \frac{\partial L}{\partial {}_aD_t^\alpha x}{}_aD_t^\alpha \eta \right) dt.$$

The above equation is also called the variation of $J[x]$ along η. For $j(\epsilon)$ to have an extremum it is necessary that $\frac{dj}{d\epsilon}\Big|_{\epsilon=0} = 0$, and this should be true for any admissible η. Thus,

$$\int_a^b \left(\frac{\partial L}{\partial x}\eta + \frac{\partial L}{\partial {}_aD_t^\alpha x}{}_aD_t^\alpha \eta \right) dt = 0$$

for all η admissible. Using the formula of integration by parts on the second term, one has

$$\int_a^b \left[\frac{\partial L}{\partial x} + {}_tD_b^\alpha \left(\frac{\partial L}{\partial {}_aD_t^\alpha x} \right) \right] \eta \, dt = 0$$

for all η admissible. The result follows by the fundamental lemma of the calculus of variations, since η is arbitrary and L is continuous. $\qquad\square$

Generalizing Theorem 4.1 for the case when L depends on several functions, i.e., $\mathbf{x}(t) = (x_1(t), \ldots, x_n(t))$ or it includes derivatives of different orders, i.e.,

$$D^\alpha \mathbf{x}(t) = (D^{\alpha_1} x_1(t), \ldots, D^{\alpha_n} x_n(t)),$$

is straightforward.

4.4 Infinite horizon fractional variational problems

Infinite horizon variational problems are a very important issue, e.g., in economics and biology, where we are interested to study a phenomena that spreads along time and not only on a finite interval $[a, b]$. We consider improper integrals with dependence on the Caputo fractional derivative, extending the main result of (Almeida and Malinowska, 2013) by considering variational problems with higher-order fractional derivatives and the fractional optimal control case. We refer the reader to Section 3.3.4 of (Malinowska and Torres, 2012).

4.4.1 The Euler–Lagrange equation

The first problem that we address is the higher-order variational problem. Let $m \in \mathbb{N}$, and for each $n \in \{1, 2, \ldots, m\}$, $\alpha_n \in (n-1, n)$ be a real. For simplification, we denote by $[x](t)$ the vector

$$[x](t) = \left(t, x(t), {}^C_a D_t^{\alpha_1} x(t), {}^C_a D_t^{\alpha_2} x(t), \ldots, {}^C_a D_t^{\alpha_m} x(t)\right)$$

applied to the function x of class $C^m[a, +\infty)$. The problem is the following: maximize the functional

$$J[x] = \int_a^{+\infty} L[x](t)\,dt, \quad \text{for } x \in C^m[a, +\infty). \tag{4.4}$$

We are assuming that the function $L : [a, +\infty) \times \mathbb{R}^{m+1} \to \mathbb{R}$ is of class C^m. By $\partial_i L[x](t)$ we mean the partial derivative of L with respect to the ith variable, evaluated at $[x](t)$. We also assume that ${}_t D_{T'}^{\alpha_n}(\partial_{n+2} L[x](t))$ is in $C[a; \infty)$, for all $n \in \{1, \ldots, m\}$.

Since we may have some problems with convergence of the integral in (4.4), we must be careful when defining maximizer trajectories for J. We follow the most common definition in the literature (Brock, 1970).

Definition 4.1. *We say that x is a weakly maximizer to the functional* (4.4) *if*

$$\lim_{T \to +\infty} \inf_{T' \geq T} \int_a^{T'} \left[L[\overline{x}](t) - L[x](t)\right] dt \leq 0$$

for all $\overline{x}$.

In the sequel we use the following lemma of (Almeida and Malinowska, 2013).

Lemma 4.1. *If g is continuous on $[a, +\infty)$ and*

$$\lim_{T \to +\infty} \inf_{T' \geq T} \int_a^{T'} g(t)h(t)\,dt = 0$$

for all continuous functions $h : [a, +\infty) \to \mathbb{R}$, then $g(t) = 0$ for all $t \geq a$.

We remark that the lemma still holds if boundary conditions are imposed on the initial time $t = a$.

Theorem 4.2. *Let x be a weakly maximizer to the functional J as in* (4.4), *subject to the initial conditions*

$$x(a) = x_a, \ x^{(n)}(a) = x_a^n, \ \text{for all } n \in \{1, 2, \ldots, m-1\}.$$

Given a function $\eta \in C^m[a, +\infty)$ and $\epsilon \in \mathbb{R}$, define

$$A(\epsilon, T') = \int_a^{T'} \frac{L[x + \epsilon\eta](t) - L[x](t)}{\epsilon} dt;$$

$$V(\epsilon, T) = \inf_{T' \geq T} \int_a^{T'} [L[x + \epsilon\eta](t) - L[x](t)] dt;$$

$$V(\epsilon) = \lim_{T \to +\infty} V(\epsilon, T).$$

Suppose that

(1) $\lim_{\epsilon \to 0} \dfrac{V(\epsilon, T)}{\epsilon}$ *exists for all T;*

(2) $\lim_{T \to +\infty} \dfrac{V(\epsilon, T)}{\epsilon}$ *exists uniformly for ϵ;*

(3) *For every $T' > a$, $T > a$, and $\epsilon \in \mathbb{R} \setminus \{0\}$, there exists a sequence $(A(\epsilon, T'_n))_{n \in \mathbb{N}}$ such that*

$$\lim_{n \to +\infty} A(\epsilon, T'_n) = \inf_{T' \geq T} A(\epsilon, T')$$

uniformly for ϵ.

Then x is a solution to the fractional Euler–Lagrange equation

$$\partial_2 L[x](t) + \sum_{n=1}^{m} {}_t D_{T'}^{\alpha_n}(\partial_{n+2} L[x](t)) = 0$$

for all $t \geq a$ and for all $T' > t$, and satisfies the transversality conditions

$$\lim_{T \to +\infty} \inf_{T' \geq T} \sum_{j=n}^{m} {}_t D_{T'}^{\alpha_j - n}(\partial_{j+2} L[x](t)) = 0 \quad at \ t = T'$$

for all $n \in \{1, \ldots, m\}$.

Proof. Observe that V is a non-positive function and $V(0) = 0$, and so $V'(0) = 0$.

Using the hypothesis of the theorem, we deduce the following:

$$V'(0) = \lim_{\varepsilon \to 0} \frac{V(\varepsilon)}{\varepsilon} = \lim_{\varepsilon \to 0} \lim_{T \to +\infty} \frac{V(\varepsilon, T)}{\varepsilon}$$

$$= \lim_{T \to +\infty} \lim_{\varepsilon \to 0} \frac{V(\varepsilon, T)}{\varepsilon}$$

$$= \lim_{T \to +\infty} \lim_{\varepsilon \to 0} \inf_{T' \geq T} A(\varepsilon, T')$$

$$= \lim_{T \to +\infty} \lim_{\varepsilon \to 0} \lim_{n \to +\infty} A(\varepsilon, T'_n)$$

$$= \lim_{T \to +\infty} \lim_{n \to +\infty} \lim_{\varepsilon \to 0} A(\varepsilon, T'_n)$$

$$= \lim_{T \to +\infty} \inf_{T' \geq T} \lim_{\varepsilon \to 0} A(\varepsilon, T')$$

$$= \lim_{T \to +\infty} \inf_{T' \geq T} \lim_{\varepsilon \to 0} \int_a^{T'} \frac{L[x + \epsilon\eta](t) - L[x](t)}{\varepsilon} \, dt$$

$$= \lim_{T \to +\infty} \inf_{T' \geq T} \int_a^{T'} \lim_{\varepsilon \to 0} \frac{L[x + \epsilon\eta](t) - L[x](t)}{\varepsilon} \, dt$$

$$= \lim_{T \to +\infty} \inf_{T' \geq T} \int_a^{T'} \left(\partial_2 L[x](t)\eta(t) + \sum_{n=1}^{m} \partial_{n+2} L[x](t)\, {}_a^C D_t^{\alpha_n} \eta(t) \right) dt.$$

Now, using the fractional integration by parts formula and rearranging terms, we get

$$0 = \lim_{T \to +\infty} \inf_{T' \geq T} \left[\int_a^{T'} \left(\partial_2 L[x](t) + \sum_{n=1}^{m} {}_t D_{T'}^{\alpha_n} (\partial_{n+2} L[x](t)) \right) \eta(t)\, dt \right.$$

$$\left. + \left[\sum_{n=1}^{m} \sum_{j=0}^{n-1} {}_t D_{T'}^{\alpha_n + j - n} (\partial_{n+2} L[x](t)) \eta^{(n-1-j)}(t) \right]_a^{T'} \right]$$

$$= \lim_{T \to +\infty} \inf_{T' \geq T} \left[\int_a^{T'} \left(\partial_2 L[x](t) + \sum_{n=1}^{m} {}_t D_{T'}^{\alpha_n} (\partial_{n+2} L[x](t)) \right) \eta(t)\, dt \right.$$

$$\left. + \left[\sum_{n=1}^{m} \eta^{(n-1)}(t) \sum_{j=n}^{m} {}_t D_{T'}^{\alpha_j - n} (\partial_{j+2} L[x](t)) \right]_a^{T'} \right].$$

We recall that by the boundary conditions imposed on the problem, the variations η are such that $\eta^{(n-1)}(a) = 0$, for all $n = 1, \ldots, m$. In order to obtain the Euler–Lagrange equation, we restrict to those variations such that $\eta^{(n-1)}(T') = 0$, for all $n = 1, \ldots, m$, and get

$$\lim_{T \to +\infty} \inf_{T' \geq T} \int_a^{T'} \left(\partial_2 L[x](t) + \sum_{n=1}^{m} {}_t D_{T'}^{\alpha_n} (\partial_{n+2} L[x](t)) \right) \eta(t)\, dt = 0.$$

Therefore, by Lemma 4.1, we have

$$\partial_2 L[x](t) + \sum_{n=1}^{m} {}_tD_{T'}^{\alpha_n}(\partial_{n+2}L[x](t)) = 0, \quad \forall t \geq a, \forall T' > t.$$

Thus, it follows that

$$\lim_{T\to+\infty} \inf_{T'\geq T} \left[\sum_{n=1}^{m} \eta^{(n-1)}(t) \sum_{j=n}^{m} {}_tD_{T'}^{\alpha_j-n}(\partial_{j+2}L[x](t)) \right]^{T'} = 0.$$

Consider variations of the kind $\eta \equiv 0$ on $[a, a+r]$ and $\eta \equiv \text{const} \neq 0$ on $[a+2r, +\infty)$, for some fixed $r > 0$. For such variations, we obtain

$$\lim_{T\to+\infty} \inf_{T'\geq T} \left[\sum_{j=1}^{m} {}_tD_{T'}^{\alpha_j-1}(\partial_{j+2}L[x](t)) \right]^{T'} = 0.$$

Repeating the process for appropriate variations, we deduce the remaining transversality conditions. □

The general case, where no constraints are imposed over the initial time $t = a$, is considered next. The proof is similar to the one of Theorem 4.2, but now the variations are such that $\eta^{(n)}(a)$ may take any value, for all $n \in \{0, 1, \ldots, m-1\}$. For such reason, we obtain m transversality conditions evaluated at $t = a$.

Theorem 4.3. *Let x be a weakly maximizer to the functional J as in (4.4). Let $A(\epsilon, T'), V(\epsilon, T), V(\epsilon)$ be as defined in Theorem 4.2, and fulfilling the same assumptions of Theorem 4.2. Then x is a solution to the fractional Euler–Lagrange equation*

$$\partial_2 L[x](t) + \sum_{n=1}^{m} {}_tD_{T'}^{\alpha_n}(\partial_{n+2}L[x](t)) = 0$$

for all $t \geq a$ and for all $T' > t$, and satisfies the transversality conditions

$$\lim_{T\to+\infty} \inf_{T'\geq T} \sum_{j=n}^{m} {}_tD_{T'}^{\alpha_j-n}(\partial_{j+2}L[x](t)) = 0 \quad \text{at } t = a \text{ and } t = T'$$

for all $n \in \{1, \ldots, m\}$.

4.4.2 *Optimal control problem*

We now consider a more general case, when the Lagrangian depends on a control given by $\frac{dz}{dt} = g(t)$ with the initial value $z(a) = 0$. Let $\alpha \in (0,1)$ and define

$$[x](t) = (t, x(t), {}^C_a D^\alpha_t x(t), z(t)) \quad \text{and} \quad \{x\}(t) = (t, x(t), {}^C_a D^\alpha_t x(t)),$$

where

$$z(t) = \int_a^t g\{x\}(\tau)\, d\tau.$$

Assume that $L : [a, +\infty) \times \mathbb{R}^3 \to \mathbb{R}$ is a function of class C^2 and $g : [a, +\infty) \times \mathbb{R}^2 \to \mathbb{R}$ is of class C^1. The objective is to determine conditions for a maximizer of the functional

$$J[x] = \int_a^{+\infty} L[x](t)dt, \quad \text{for } x \in C^1[a, +\infty), \tag{4.5}$$

where we understand maximizers in the sense of Definition 4.1 with the adequate adjustments.

Theorem 4.4. *Let x be a weakly maximizer to the functional J as in (4.5), subject to the initial condition $x(a) = x_a$. Given a function $\eta \in C^1[a, +\infty)$ and $\epsilon \in \mathbb{R}$, define*

$$A(\epsilon, T') = \int_a^{T'} \frac{L[x + \epsilon\eta](t) - L[x](t)}{\epsilon} dt;$$

$$V(\epsilon, T) = \inf_{T' \geq T} \int_a^{T'} [L[x + \epsilon\eta](t) - L[x](t)]dt;$$

$$V(\epsilon) = \lim_{T \to +\infty} V(\epsilon, T).$$

Suppose that

(1) $\lim\limits_{\epsilon \to 0} \dfrac{V(\epsilon, T)}{\epsilon}$ *exists for all T;*

(2) $\lim\limits_{T \to +\infty} \dfrac{V(\epsilon, T)}{\epsilon}$ *exists uniformly for ϵ;*

(3) For every $T' > a$, $T > a$, and $\epsilon \in \mathbb{R} \setminus \{0\}$, there exists a sequence $(A(\epsilon, T'_n))_{n \in \mathbb{N}}$ such that

$$\lim_{n \to +\infty} A(\epsilon, T'_n) = \inf_{T' \geq T} A(\epsilon, T')$$

uniformly for ϵ.

Then x is a solution to the fractional Euler–Lagrange equation

$$\partial_2 L[x](t) + {}_tD_{T'}^\alpha(\partial_3 L[x](t)) + \left(\int_t^{T'} \partial_4 L[x](\tau)\, d\tau\right) \partial_2 g\{x\}(t)$$

$$+ {}_tD_{T'}^\alpha\left(\left(\int_t^{T'} \partial_4 L[x](\tau)\, d\tau\right) \partial_3 g\{x\}(t)\right) = 0$$

for all $t \geq a$ and for all $T' > t$, and satisfies the transversality condition

$$\lim_{T\to+\infty} \inf_{T'\geq T} {}_tI_{T'}^{1-\alpha}\left(\partial_3 L[x](t) + \left(\int_t^{T'} \partial_4 L[x](\tau)\, d\tau\right) \partial_3 g\{x\}(t)\right) = 0$$

at $t = T'$.

Proof. Following similar arguments as the ones exemplified in the proof of Theorem 4.2, we arrive to

$$\lim_{T\to+\infty} \inf_{T'\geq T} \int_a^{T'} \lim_{\varepsilon\to 0} \frac{L[x+\varepsilon\eta](t) - L[x](t)}{\varepsilon}\, dt = 0,$$

that is,

$$\lim_{T\to+\infty} \inf_{T'\geq T} \int_a^{T'} \left[\partial_2 L[x](t)\eta(t) + \partial_3 L[x](t){}_a^C D_t^\alpha \eta(t)\right.$$

$$\left. + \partial_4 L[x](t)\int_a^t \left(\partial_2 g\{x\}(\tau)\eta(\tau) + \partial_3 g\{x\}(\tau){}_a^C D_t^\alpha \eta(\tau)\right) d\tau\right] dt = 0.$$

Using the relations
R1:

$$\int_a^{T'} \partial_3 L[x](t){}_a^C D_t^\alpha \eta(t)\, dt$$

$$= \int_a^{T'} {}_tD_{T'}^\alpha(\partial_3 L[x](t))\eta(t)\, dt + \left[{}_tI_{T'}^{1-\alpha}(\partial_3 L[x](t))\eta(t)\right]_a^{T'};$$

R2:

$$\int_a^{T'} \partial_4 L[x](t) \left(\int_a^t \partial_2 g\{x\}(\tau)\eta(\tau)\,d\tau \right) dt$$

$$= \int_a^{T'} \left(-\frac{d}{dt} \int_t^{T'} \partial_4 L[x](\tau)\,d\tau \right) \left(\int_a^t \partial_2 g\{x\}(\tau)\eta(\tau)\,d\tau \right) dt$$

$$= \int_a^{T'} \left(\int_t^{T'} \partial_4 L[x](\tau)\,d\tau \right) \partial_2 g\{x\}(t)\eta(t)\,dt$$

$$- \left[\left(\int_t^{T'} \partial_4 L[x](\tau)\,d\tau \right) \left(\int_a^t \partial_2 g\{x\}(\tau)\eta(\tau)\,d\tau \right) \right]_a^{T'}$$

$$= \int_a^{T'} \left(\int_t^{T'} \partial_4 L[x](\tau)\,d\tau \right) \partial_2 g\{x\}(t)\eta(t)\,dt;$$

R3:

$$\int_a^{T'} \partial_4 L[x](t) \left(\int_a^t \partial_3 g\{x\}(\tau)\,_a^C D_\tau^\alpha \eta(\tau)\,d\tau \right) dt$$

$$= \int_a^{T'} \left(-\frac{d}{dt} \int_t^{T'} \partial_4 L[x](\tau)\,d\tau \right) \left(\int_a^t \partial_3 g\{x\}(\tau)\,_a^C D_\tau^\alpha \eta(\tau)\,d\tau \right) dt$$

$$= \int_a^{T'} \left(\int_t^{T'} \partial_4 L[x](\tau)\,d\tau \right) \partial_3 g\{x\}(t)\,_a^C D_t^\alpha \eta(t)\,dt$$

$$- \left[\left(\int_t^{T'} \partial_4 L[x](\tau)\,d\tau \right) \left(\int_a^t \partial_3 g\{x\}(\tau)\,_a^C D_\tau^\alpha \eta(\tau)\,d\tau \right) \right]_a^{T'}$$

$$= \int_a^{T'} {}_t D_{T'}^\alpha \left(\left(\int_t^{T'} \partial_4 L[x](\tau)\,d\tau \right) \partial_3 g\{x\}(t) \right) \eta(t)\,dt$$

$$+ \left[{}_t I_{T'}^{1-\alpha} \left(\left(\int_t^{T'} \partial_4 L[x](\tau)\,d\tau \right) \partial_3 g\{x\}(t) \right) \eta(t) \right]_a^{T'};$$

we obtain

$$\lim_{T\to+\infty} \inf_{T'\geq T} \left\{ \int_a^{T'} \left[\partial_2 L[x](t) + {}_t D_{T'}^\alpha(\partial_3 L[x](t)) + \left(\int_t^{T'} \partial_4 L[x](\tau)\,d\tau \right) \right. \right.$$

$$\times \partial_2 g\{x\}(t) + {}_t D_{T'}^\alpha \left(\left(\int_t^{T'} \partial_4 L[x](\tau)\,d\tau \right) \partial_3 g\{x\}(t) \right) \bigg] \eta(t)\,dt$$

$$\left. + \left[{}_t I_{T'}^{1-\alpha} \left(\partial_3 L[x](t) + \left(\int_t^{T'} \partial_4 L[x](\tau)\,d\tau \right) \partial_3 g\{x\}(t) \right) \eta(t) \right]_a^{T'} \right\} = 0.$$

Since $\eta(a) = 0$, by the boundary condition at $t = a$, and restricting to variations such that $\eta(T') = 0$, we deduce by Lemma 4.1 the fractional Euler–Lagrange equation. Therefore, the transversality condition must be also satisfied. □

Theorem 4.5. *Let x be a weakly maximizer to the functional J as in (4.5). Let $A(\epsilon, T'), V(\epsilon, T), V(\epsilon)$ be as defined in Theorem 4.4, fulfilling the same assumptions. Then x is a solution to the fractional Euler–Lagrange equation*

$$\partial_2 L[x](t) + {}_tD_{T'}^\alpha (\partial_3 L[x](t)) + \left(\int_t^{T'} \partial_4 L[x](\tau)\, d\tau\right) \partial_2 g\{x\}(t)$$

$$+ {}_tD_{T'}^\alpha \left(\left(\int_t^{T'} \partial_4 L[x](\tau)\, d\tau\right) \partial_3 g\{x\}(t)\right) = 0$$

for all $t \geq a$ and for all $T' > t$, and satisfies the transversality condition

$$\lim_{T\to+\infty} \inf_{T'\geq T} {}_tI_{T'}^{1-\alpha} \left(\partial_3 L[x](t) + \left(\int_t^{T'} \partial_4 L[x](\tau)\, d\tau\right) \partial_3 g\{x\}(t)\right) = 0$$

at $t = a$ and $t = T'$.

4.4.3 *Example*

We provide an example of application of Theorem 4.2. Let

$$J[x] = \int_0^{+\infty} \left[t^2 - x^3 + ({}_0^C D_t^{1/2} x(t))^3 + 3({}_0^C D_t^{3/2} x(t))^2\right] dt$$

subject to the boundary conditions $x(0) = x'(0) = 0$. The Euler–Lagrange equation associated to it is

$$x^2(t) = {}_tD_{T'}^{1/2} ({}_0^C D_t^{1/2} x(t))^2 + 2{}_tD_{T'}^{3/2} ({}_0^C D_t^{3/2} x(t)) = 0$$

for all $t \geq 0$ and for all $T' > t$, and the two transversality conditions are

$$\lim_{T\to+\infty} \inf_{T'\geq T} \left[{}_tI_{T'}^{1/2} ({}_0^C D_t^{1/2} x(t))^2 + 2{}_tD_{T'}^{1/2} ({}_0^C D_t^{3/2} x(t))\right] = 0$$

and

$$\lim_{T\to+\infty} \inf_{T'\geq T} \left[{}_tD_{T'}^{1/2} ({}_0^C D_t^{3/2} x(t))\right] = 0$$

at $t = T'$.

4.5   Variational problems with the Riesz–Caputo derivative

The aim of this section is to investigate optimality conditions for fractional variational problems with a Lagrangian depending on the Riesz–Caputo derivative. We consider several cases: when the interval of integration of the functional is different from the interval of the fractional derivative, problems in presence of integral dynamic constraints, and functionals depending also on the terminal time.

4.5.1   *The Euler–Lagrange equation*

We start by recalling the Riesz–Caputo fractional derivative ${}_{a}^{RC}D_{b}^{\alpha}x(t)$ and the Riesz fractional derivative ${}_{a}^{R}D_{b}^{\alpha}x(t)$. Given a function x, the Riesz–Caputo fractional derivative is defined as

$$
{}_{a}^{RC}D_{b}^{\alpha}x(t) = \frac{1}{2}({}_{a}^{C}D_{t}^{\alpha}x(t) - {}_{t}^{C}D_{b}^{\alpha}x(t)).
$$

The Riesz fractional derivative ${}_{a}^{R}D_{b}^{\alpha}x(t)$ is given by

$$
{}_{a}^{R}D_{b}^{\alpha}x(t) = \frac{1}{2}({}_{a}D_{t}^{\alpha}x(t) - {}_{t}D_{b}^{\alpha}x(t)).
$$

It is easy to verify that the integration by parts formula reads as

$$
\int_{a}^{b} x(t) \cdot {}_{a}^{RC}D_{b}^{\alpha}y(t)dt = -\int_{a}^{b} {}_{a}^{R}D_{b}^{\alpha}x(t) \cdot y(t)dt + {}_{a}^{R}I_{b}^{1-\alpha}x(t) \cdot y(t)\Big|_{a}^{b},
$$

where

$$
{}_{a}^{R}I_{b}^{\alpha}x(t) = \frac{1}{2}({}_{a}I_{t}^{\alpha}x(t) + {}_{t}I_{b}^{\alpha}x(t))
$$

is called the Riesz fractional integral.

The fundamental fractional variational problem that we address here is stated in the following way.

(P1) Among all C^{1} functions $x : [a, b] \to \mathbb{R}$, with fixed values on $t = a$ and $t = b$, say

$$
x(a) = x_{a} \text{ and } x(b) = x_{b}, \quad x_{a}, x_{b} \in \mathbb{R},
$$

find the ones for which the functional

$$
J[x] = \int_{a}^{b} L\left(t, x(t), {}_{a}^{RC}D_{b}^{\alpha}x(t)\right) dt
$$

attains a minimum value.

We are assuming, here and from now on, that the Lagrange function $L : [a, b] \times \mathbb{R}^{2} \to \mathbb{R}$ is a function with continuous first and second partial derivatives with respect to all its arguments, and $_{a}^{RC}D_{b}^{\alpha}x(t)$ exists and is continuous on the closed interval $[a, b]$. Also, by $[x](t)$ we mean

$$[x](t) = \left(t, x(t), {}_{a}^{RC}D_{b}^{\alpha}x(t)\right).$$

The possible extremizers for problem (P1) can be obtained by solving a fractional differential equation, the so called fractional Euler–Lagrange equation.

Theorem 4.6 (See (Agrawal, 2007a)). *Let x be a solution to problem (P1). Then, x is a solution of the fractional Euler–Lagrange equation*

$$\frac{\partial L}{\partial x}[x](t) - {}^{R}_{a}D_{b}^{\alpha}\frac{\partial L}{\partial {}_{a}^{RC}D_{b}^{\alpha}x}[x](t) = 0 \tag{4.6}$$

for all $t \in [a, b]$.

We remark that, when $\alpha = 1$, Eq. (4.6) is the standard Euler–Lagrange equation: if x is a minimizer or maximizer of

$$J[x] = \int_{a}^{b} L(t, x(t), x^{(1)}(t))dt$$

subject to the boundary conditions

$$x(a) = x_{a} \text{ and } x(b) = x_{b},$$

then x is a solution of the differential equation

$$\frac{\partial L}{\partial x}(t, x(t), x^{(1)}(t)) - \frac{d}{dt}\frac{\partial L}{\partial x^{(1)}}(t, x(t), x^{(1)}(t)) = 0.$$

Definition 4.2. *A function x that is a solution of Eq. (4.6) is called an extremal for J.*

We first extend Theorem 4.6 to functionals where the interval of integration is $[A, B] \subset [a, b]$. The idea comes from (Almeida and Torres, 2011; Atanacković, Konjik and Pilipović, 2008), where similar problems with a Lagrangian function depending on the Riemann–Liouville (Atanacković, Konjik and Pilipović, 2008) or the Caputo (Almeida and Torres, 2011) fractional derivatives are considered.

The new problem (P2) deals with finding optimality conditions for functionals of type

$$J[x] = \int_{A}^{B} L\left(t, x(t), {}_{a}^{RC}D_{b}^{\alpha}x(t)\right) dt$$

with boundary conditions
$$x(a) = x_a \text{ and } x(b) = x_b, \quad x_a, x_b \in \mathbb{R}.$$

Theorem 4.7. *Let x be a solution to problem (P2). Then, x satisfies the following equations:*

$$
\begin{cases}
\dfrac{\partial L}{\partial x}[x](t) - {}^{R}_{A}D^{\alpha}_{B}\dfrac{\partial L}{\partial^{RC}_{a}D^{\alpha}_{b}x}[x](t) = 0, & \text{for all } t \in [A, B], \\[2ex]
{}_{t}D^{\alpha}_{B}\dfrac{\partial L}{\partial^{RC}_{a}D^{\alpha}_{b}x}[x](t) - {}_{t}D^{\alpha}_{A}\dfrac{\partial L}{\partial^{RC}_{a}D^{\alpha}_{b}x}[x](t) = 0, & \text{for all } t \in [a, A], \\[2ex]
{}_{B}D^{\alpha}_{t}\dfrac{\partial L}{\partial^{RC}_{a}D^{\alpha}_{b}x}[x](t) - {}_{A}D^{\alpha}_{t}\dfrac{\partial L}{\partial^{RC}_{a}D^{\alpha}_{b}x}[x](t) = 0, & \text{for all } t \in [B, b].
\end{cases}
$$

Proof. To obtain the necessary conditions, we first consider variation functions of the type $x + \epsilon h$, where $h : [a, b] \to \mathbb{R}$ is a function of class C^1 such that $h(a) = h(b) = 0$. For convenience, we also assume that $h(A) = h(B) = 0$. Let $j(\epsilon) = J[x + \epsilon h]$. Since $j'(0) = 0$, integrating by parts, we obtain

$$
0 = \int_A^B \left[\frac{\partial L}{\partial x}[x](t)h(t) + \frac{\partial L}{\partial^{RC}_{a}D^{\alpha}_{b}x}[x](t){}^{RC}_{a}D^{\alpha}_{b}h(t) \right] dt
$$

$$
= \int_A^B \frac{\partial L}{\partial x}[x](t)h(t)\, dt
$$

$$
+ \frac{1}{2}\left[\int_A^B \frac{\partial L}{\partial^{RC}_{a}D^{\alpha}_{b}x}[x](t){}^{C}_{a}D^{\alpha}_{t}h(t)\, dt - \int_a^A \frac{\partial L}{\partial^{RC}_{a}D^{\alpha}_{b}x}[x](t){}^{C}_{a}D^{\alpha}_{t}h(t)\, dt \right]
$$

$$
- \frac{1}{2}\left[\int_A^b \frac{\partial L}{\partial^{RC}_{a}D^{\alpha}_{b}x}[x](t){}^{C}_{t}D^{\alpha}_{b}h(t)\, dt - \int_B^b \frac{\partial L}{\partial^{RC}_{a}D^{\alpha}_{b}x}[x](t){}^{C}_{t}D^{\alpha}_{b}h(t)\, dt \right]
$$

$$
= \int_A^B \frac{\partial L}{\partial x}[x](t)h(t)\, dt
$$

$$
+ \frac{1}{2}\left[\int_u^B {}_{t}D^{\alpha}_{B}\frac{\partial L}{\partial^{RC}_{a}D^{\alpha}_{b}x}[x](t)h(t)\, dt - \int_a^A {}_{t}D^{\alpha}_{A}\frac{\partial L}{\partial^{RC}_{a}D^{\alpha}_{b}x}[x](t)h(t)\, dt \right]
$$

$$
- \frac{1}{2}\left[\int_A^b {}_{A}D^{\alpha}_{t}\frac{\partial L}{\partial^{RC}_{a}D^{\alpha}_{b}x}[x](t)h(t)\, dt - \int_B^b {}_{B}D^{\alpha}_{t}\frac{\partial L}{\partial^{RC}_{a}D^{\alpha}_{b}x}[x](t)h(t)\, dt \right]
$$

$$
= \int_A^B \left[\frac{\partial L}{\partial x}[x](t) - {}^{R}_{A}D^{\alpha}_{B}\frac{\partial L}{\partial^{RC}_{a}D^{\alpha}_{b}x}[x](t) \right] h(t)\, dt
$$

$$
+ \frac{1}{2}\int_a^A \left[{}_{t}D^{\alpha}_{B}\frac{\partial L}{\partial^{RC}_{a}D^{\alpha}_{b}x}[x](t) - {}_{t}D^{\alpha}_{A}\frac{\partial L}{\partial^{RC}_{a}D^{\alpha}_{b}x}[x](t) \right] h(t)\, dt
$$

$$
- \frac{1}{2}\int_B^b \left[{}_{B}D^{\alpha}_{t}\frac{\partial L}{\partial^{RC}_{a}D^{\alpha}_{b}x}[x](t) - {}_{A}D^{\alpha}_{t}\frac{\partial L}{\partial^{RC}_{a}D^{\alpha}_{b}x}[x](t) \right] h(t)\, dt.
$$

For appropriate choices of h, we prove the necessary conditions. □

Obviously, in the case $A = a$ and $B = b$, Theorem 4.7 reduces to Theorem 4.6. Other cases could be deduced from Theorem 4.7, namely when $A = a$ and $B \neq b$, or $A \neq a$ and $B = b$.

4.5.2 *The fractional isoperimetric problem*

The original isoperimetric problem is addressed in the following way: among all closed plane curves, without self-intersecting, such that the total length has a given value, find the ones for which the enclosed area is the greatest. Nowadays, isoperimetric problems are the ones that involve some integral constraint on the dynamics, and have become one of the classical problems of the calculus of variations. The fractional isoperimetric problem is stated as follows.

(P3) Find a function $x : [a, b] \to \mathbb{R}$ of class C^1 such that it minimizes the functional

$$J[x] = \int_a^b L(t, x(t), {}_a^{RC}D_b^{\alpha}x(t))dt$$

subject to the boundary conditions

$$x(a) = x_a \text{ and } x(b) = x_b, \quad x_a, x_b \in \mathbb{R},$$

and to an integral constraint

$$I[x] = \int_a^b g(t, x(t), {}_a^{RC}D_b^{\alpha}x(t))dt = l,$$

where l is a fixed real. As before, we assume that $g : [a, b] \times \mathbb{R}^2 \to \mathbb{R}$ is a function with continuous first and second partial derivatives with respect to all its arguments.

Theorem 4.8. *Let x be a solution to problem (P3). If x is not an extremal for I, then there exists a constant λ such that x satisfies the equation*

$$\frac{\partial F}{\partial x}[x](t) - {}_a^R D_b^{\alpha} \frac{\partial F}{\partial {}_a^{RC}D_b^{\alpha}x}[x](t) = 0$$

for all $t \in [a, b]$, where $F = L - \lambda g$.

Proof. Consider variations of x of the form

$$x + \epsilon_1 h_1 + \epsilon_2 h_2,$$

where h_i is a function of class C^1, and $h_i(a) = h_i(b) = 0$, for each $i \in \{1,2\}$. Define two functions j and i by

$$j(\epsilon_1, \epsilon_2) = J[x + \epsilon_1 h_1 + \epsilon_2 h_2] \text{ and } i(\epsilon_1, \epsilon_2) = I[x + \epsilon_1 h_1 + \epsilon_2 h_2] - l.$$

Since x is not an extremal for I, there exists a function h_2 for which

$$\left[\frac{\partial i}{\partial \epsilon_2}\right]_{(0,0)} \neq 0.$$

Also, since $i(0,0) = 0$, by the implicit function theorem, there exists a function ϵ_2 satisfying the relation $i(\epsilon_1, \epsilon_2(\epsilon_1)) = 0$. In other words, there exists a subfamily of variation functions satisfying the integral constraint. Moreover, observe that j has a minimum at zero subject to the constraint $i(\epsilon_1, \epsilon_2) = 0$, and we just proved that $\nabla i(0,0) \neq (0,0)$. Then, by the Lagrange multiplier rule, there exists a constant λ such that

$$\nabla(j(0,0) - \lambda i(0,0)) = (0,0).$$

Differentiating j and i with respect to ϵ_1, at $(\epsilon_1, \epsilon_2) = (0,0)$, we prove the theorem. $\qquad\square$

We now shall present a more general result.

Theorem 4.9. *Let x be solution to problem (P3). Then there exist two constants λ_0 and λ, not both zero, such that y satisfies the equation*

$$\frac{\partial K}{\partial x}[x](t) - {}_a^R D_b^\alpha \frac{\partial K}{\partial {}_a^{RC} D_b^\alpha x}[x](t) = 0$$

for all $t \in [a,b]$, where $K = \lambda_0 L - \lambda g$.

Proof. It follows the same pattern as the proof of Theorem 4.8, and using the abnormal Lagrange multiplier rule. $\qquad\square$

The isoperimetric problem for functionals where the interval of integration is $[A, B] \subset [a,b]$ can be solved in a similar way. We state problem (P4) as follows:

$$\text{minimize } J[x] = \int_A^B L(t, x(t), {}_a^{RC} D_b^\alpha x(t)) dt$$

subject to the boundary conditions

$$x(a) = x_a \text{ and } x(b) = x_b$$

and to an integral constraint

$$I[x] = \int_A^B g(t, x(t), {}_a^{RC} D_b^\alpha x(t)) dt = l.$$

Similarly, we say that x is an extremal for I if

$$\frac{\partial L}{\partial x}[x](t) - {}_A^C D_B^\alpha \frac{\partial L}{\partial {}_a^{RC} D_b^\alpha x}[x](t) = 0$$

for all $t \in [A, B]$. Using the same techniques as the ones presented in Theorems 4.7 and 4.8, the following two results can be proven.

Theorem 4.10. *If x is a solution to problem (P4), and if x is not an extremal for I, then there exists a constant λ such that*

$$\begin{cases} \dfrac{\partial F}{\partial x}[x](t) - {}_A^R D_B^\alpha \dfrac{\partial F}{\partial {}_a^{RC} D_b^\alpha x}[x](t) = 0 & \text{for all } t \in [A, B], \\[2mm] {}_t D_B^\alpha \dfrac{\partial F}{\partial {}_a^{RC} D_b^\alpha x}[x](t) - {}_t D_A^\alpha \dfrac{\partial F}{\partial {}_a^{RC} D_b^\alpha x}[x](t) = 0 & \text{for all } t \in [a, A], \\[2mm] {}_B D_t^\alpha \dfrac{\partial F}{\partial {}_a^{RC} D_b^\alpha x}[x](t) - {}_A D_t^\alpha \dfrac{\partial F}{\partial {}_a^{RC} D_b^\alpha x}[x](t) = 0 & \text{for all } t \in [B, b] \end{cases}$$

with $F = L - \lambda g$.

Theorem 4.11. *If x is a solution to problem (P4), then there exist two constants λ_0 and λ, not both zero, such that*

$$\begin{cases} \dfrac{\partial K}{\partial x}[x](t) - {}_A^R D_B^\alpha \dfrac{\partial K}{\partial {}_a^{RC} D_b^\alpha x}[x](t) = 0 & \text{for all } t \in [A, B], \\[2mm] {}_t D_B^\alpha \dfrac{\partial K}{\partial {}_a^{RC} D_b^\alpha x}[x](t) - {}_t D_A^\alpha \dfrac{\partial K}{\partial {}_a^{RC} D_b^\alpha x}[x](t) = 0 & \text{for all } t \in [a, A], \\[2mm] {}_B D_t^\alpha \dfrac{\partial K}{\partial {}_a^{RC} D_b^\alpha x}[x](t) - {}_A D_t^\alpha \dfrac{\partial K}{\partial {}_a^{RC} D_b^\alpha x}[x](t) = 0 & \text{for all } t \in [B, b] \end{cases}$$

with $K = \lambda_0 L - \lambda g$.

4.5.3 Optimal time problem

Now we are interested not only in finding an optimal admissible function for the variational problem, but also the optimal time T. We state the problem in the following way: consider the functional

$$J[x, T] = \int_a^T L(t, x(t), {}_a^{RC} D_b^\alpha x(t)) dt,$$

where $(x, T) \in \{C^1[a, b] \times [a, b] \mid x(a) = x_a\}$. Problem (P5) is the following one: find a pair (x, T) for which J attains a minimum value.

Theorem 4.12. *If (x, T) is a solution to problem (P5), then it satisfies the following four conditions:*

(1) $L(T, x(T), {}_{a}^{RC}D_{b}^{\alpha}x(T)) = 0$,

(2) $\dfrac{\partial L}{\partial x}[x](t) - {}_{a}^{R}D_{T}^{\alpha}\dfrac{\partial L}{\partial {}_{a}^{RC}D_{b}^{\alpha}x}[x](t) = 0$ *for all* $t \in [a, T]$,

(3) ${}_{a}D_{t}^{\alpha}\dfrac{\partial L}{\partial {}_{a}^{RC}D_{b}^{\alpha}x}[x](t) = {}_{T}D_{t}^{\alpha}\dfrac{\partial L}{\partial {}_{a}^{RC}D_{b}^{\alpha}x}[x](t)$ *for all* $t \in [T, b]$,

(4) $\left[{}_{a}I_{t}^{1-\alpha}\dfrac{\partial L}{\partial {}_{a}^{RC}D_{b}^{\alpha}x}[x](t) \right]_{b} = \left[{}_{T}I_{t}^{1-\alpha}\dfrac{\partial L}{\partial {}_{a}^{RC}D_{b}^{\alpha}x}[x](t) \right]_{b}$.

Proof.   Let $h : [a, b] \to \mathbb{R}$ be a function of class C^1 such that $h(a) = 0$, and let $\Delta T \in \mathbb{R}$. Define j as

$$j(\epsilon) = J[x + \epsilon h, T + \epsilon \Delta T].$$

Since $j'(0) = 0$, we get

$$\int_{a}^{T} \left[\frac{\partial L}{\partial x}[x](t)h(t) + \frac{\partial L}{\partial {}_{a}^{RC}D_{b}^{\alpha}x}[x](t) {}_{a}^{RC}D_{b}^{\alpha}h(t) \right] dt$$
$$+ \Delta T \cdot L(T, x(T), {}_{a}^{RC}D_{b}^{\alpha}x(T)) = 0.$$

On the other hand, observe that

$$\int_{a}^{T} \frac{\partial L}{\partial {}_{a}^{RC}D_{b}^{\alpha}x}[x](t) {}_{a}^{RC}D_{b}^{\alpha}h(t)\, dt$$

$$= \frac{1}{2} \left[\int_{a}^{T} \frac{\partial L}{\partial {}_{a}^{RC}D_{b}^{\alpha}x}[x](t) {}_{a}^{C}D_{t}^{\alpha}h(t)\, dt - \int_{a}^{T} \frac{\partial L}{\partial {}_{a}^{RC}D_{b}^{\alpha}x}[x](t) {}_{t}^{C}D_{b}^{\alpha}h(t)\, dt \right]$$

$$= \frac{1}{2} \left[\int_{a}^{T} \frac{\partial L}{\partial {}_{a}^{RC}D_{b}^{\alpha}x}[x](t) {}_{a}^{C}D_{t}^{\alpha}h(t)\, dt - \int_{a}^{b} \frac{\partial L}{\partial {}_{a}^{RC}D_{b}^{\alpha}x}[x](t) {}_{t}^{C}D_{b}^{\alpha}h(t)\, dt \right.$$

$$\left. + \int_{T}^{b} \frac{\partial L}{\partial {}_{a}^{RC}D_{b}^{\alpha}x}[x](t) {}_{t}^{C}D_{b}^{\alpha}h(t)\, dt \right]$$

$$= \star$$

Integrating by parts each of the last three terms, we get

$$\star = \frac{1}{2} \left[\int_{a}^{T} {}_{t}D_{T}^{\alpha}\frac{\partial L}{\partial {}_{a}^{RC}D_{b}^{\alpha}x}[x](t)h(t)\, dt + \left[{}_{t}I_{T}^{1-\alpha}\frac{\partial L}{\partial {}_{a}^{RC}D_{b}^{\alpha}x}[x](t)h(t) \right]_{T} \right.$$

$$- \int_{a}^{b} {}_{a}D_{t}^{\alpha}\frac{\partial L}{\partial {}_{a}^{RC}D_{b}^{\alpha}x}[x](t)h(t)\, dt + \left[{}_{a}I_{t}^{1-\alpha}\frac{\partial L}{\partial {}_{a}^{RC}D_{b}^{\alpha}x}[x](t)h(t) \right]_{b}$$

$$\left. + \int_{T}^{b} {}_{T}D_{t}^{\alpha}\frac{\partial L}{\partial {}_{a}^{RC}D_{b}^{\alpha}x}[x](t)h(t)\, dt + \left[{}_{T}I_{t}^{1-\alpha}\frac{\partial L}{\partial {}_{a}^{RC}D_{b}^{\alpha}x}[x](t)h(t) \right]_{b}^{T} \right].$$

Some of the previous terms vanish (cf. (Miller and Ross, 1993, p. 46)):

$$\left[{}_t I_T^{1-\alpha} \frac{\partial L}{\partial_a^{RC} D_b^\alpha x}[x](t)h(t) \right]_T = 0 \text{ and } \left[{}_T I_t^{1-\alpha} \frac{\partial L}{\partial_a^{RC} D_b^\alpha x}[x](t)h(t) \right]_T = 0.$$

Thus, we have proven the relation

$$0 = \int_a^T \left[\frac{\partial L}{\partial x}[x](t) - {}_a^R D_T^\alpha \frac{\partial L}{\partial_a^{RC} D_b^\alpha x}[x](t) \right] h(t)\, dt$$

$$+ \Delta T \cdot L(T, x(T), {}_a^{RC} D_b^\alpha x(T))$$

$$+ \frac{1}{2} \int_T^b \left[{}_T D_t^\alpha \frac{\partial L}{\partial_a^{RC} D_b^\alpha x}[x](t) - {}_a D_t^\alpha \frac{\partial L}{\partial_a^{RC} D_b^\alpha x}[x](t) \right] h(t)\, dt$$

$$+ \frac{1}{2} \left[\left[{}_a I_t^{1-\alpha} \frac{\partial L}{\partial_a^{RC} D_b^\alpha x}[x](t) - {}_T I_t^{1-\alpha} \frac{\partial L}{\partial_a^{RC} D_b^\alpha x}[x](t) \right] h(t) \right]_b.$$

If we fix $h \equiv 0$, by the arbitrariness of ΔT, we obtain equation (1) of the theorem. If h is free on $[a, T)$ and zero on $[T, b]$, we obtain equation (2). To prove the two remaining conditions, choose first h free on $]T, b[$ and zero on $t = b$, and then h such that $h(b) \neq 0$. □

The transversality condition obtained in (Agrawal, 2007a) can be seen as a particular case of Theorem 4.12. Let $T = b$ (time is fixed). Then $\Delta T = 0$. Following the proof, equation (1) of Theorem 4.12 is no longer a necessary condition. Equation (2) becomes the fractional Euler–Lagrange equation as in Theorem 4.6. Equation (3) is obviously satisfied. About equation (4), it reads as (cf. (Miller and Ross, 1993, p. 46))

$$\left[{}_a I_t^{1-\alpha} \frac{\partial L}{\partial_a^{RC} D_b^\alpha x}[x](t) \right]_b = 0,$$

which is equivalent to the one obtained in (Agrawal, 2007a):

$$\left[{}_a^R I_b^{1-\alpha} \frac{\partial L}{\partial_a^{RC} D_b^\alpha x}[x](t) \right]_b = 0.$$

4.6 Solution methods

There are two main approaches to solve variational, including optimal control, problems. On the one hand, there are the direct methods. In a branch of direct methods, the problem is discretized on the interested time interval using discrete values of the unknown function, finite differences for derivatives and finally a quadrature rule for the integral. This procedure transforms the variational problem, a dynamic optimization problem, to a static multivariable optimization problem. Better accuracies are achieved by refining the underlying mesh size. Another class of direct methods uses function approximation through a linear combination of the elements of a certain basis, e.g., power series. The problem is then transformed to the determination of the unknown coefficients. To get better results in this sense, is the matter of using more adequate or higher-order function approximations.

On the other hand, there are the indirect methods. Those transform a variational problem to an equivalent differential equation by applying some necessary optimality conditions. Euler–Lagrange equations and Pontryagin's minimum principle are used in this context to make the transformation process. Once we solve the resulting differential equation, an extremal for the original problem is reached. Therefore, to reach better results using indirect methods, one has to employ powerful integrators. It is worth, however, mentioning here that numerical methods are usually used to solve practical problems.

These two classes of methods have been generalized to cover fractional problems, since that is the essential subject of this book.

Chapter 5

Numerical methods for fractional variational problems

In this chapter we give a short survey on the numerical methods for solving fractional variational problems. As mentioned earlier, the fractional calculus of variations started with the works of Riewe (1996, 1997), in the last years of the nineties. Later, the notion of fractional optimal control appeared in the works of Agrawal (2004) and Frederico and Torres (2008a). It is not surprising that the numerical achievements in these fields is at an early stage. In this chapter we shall review some recent papers which can be classified as direct or indirect methods.

The first effort to solve a fractional optimal control problem numerically was made by Agrawal in 2004 (Agrawal, 2004). The problem under consideration consists in finding an optimal control u, which minimizes the functional

$$J[x, u] = \int_0^1 F\left(t, x(t), u(t)\right) dt,$$

while it is assumed to satisfy a given dynamic constraint of the form

$$_aD_t^\alpha x(t) = G\left(t, x(t), u(t)\right)$$

subject to the boundary condition

$$x(0) = x_0.$$

The Euler–Lagrange equation can be derived by using a Lagrange multiplier, λ (Frederico and Torres, 2008a). The necessary optimality condition reads to

$$\begin{cases} _aD_t^\alpha x(t) = G(t, x(t), u(t)) \\ _tD_1^\alpha \lambda(t) = \dfrac{\partial F}{\partial x} + \lambda(t)\dfrac{\partial G}{\partial x} \\ 0 = \dfrac{\partial F}{\partial u} + \lambda(t)\dfrac{\partial G}{\partial u} \end{cases}, \qquad \begin{cases} x(0) = x_0 \\ \lambda(1) = 0. \end{cases}$$

The paper (Agrawal, 2004) uses a Ritz method by approximating x and λ using shifted Legendre polynomials, i.e.,

$$x(t) \approx \sum_{j=1}^{m} c_j P_j(t), \qquad \lambda(t) \approx \sum_{j=1}^{m} c_j P_j(t).$$

The shifted Legendre polynomials are explicitly given by

$$P_n(t) = (-1)^n \sum_{k=0}^{n} \binom{n}{k} \binom{n+k}{k} (-x)^k.$$

One can use the orthogonality of Legendre polynomials and the fact that their fractional derivatives are available in closed forms. This method, after some calculus operations and simplifications, leads to a system of $2m + 2$ equations in $2m + 2$ unknowns. Approximate solutions to the problem then is achieved in terms of linear combinations of the shifted Legendre polynomials.

The same idea has been tried later by several authors. This is done by either using different approximations in terms of other basis functions or a different class of variational problems, say in the problem formulation or in the fractional term that appears.

Approximating x, u and λ by multiwavelets is an example of a new version of this method. In (Lotfi, Dehghan and Yousefi, 2011) the Caputo fractional derivative is used in the constraint and another functional is considered. Other aspects like some properties of Legendre polynomials and the convergence are also covered in this work.

Another slightly different approach is the use of the so-called multi-wavelet collocation that has been introduced in (Yousefi, Lotfi and Dehghan, 2011). The method is based on the approximations

$$x(t) \approx \sum_{i=0}^{2^k-1} \sum_{j=0}^{M} (t-a) c x_{ij} \psi_{ij}(t) + x_0,$$

$$u(t) \approx \sum_{i=0}^{2^k-1} \sum_{j=0}^{M} c u_{ij} \psi_{ij}(t),$$

$$\lambda(t) \approx \sum_{i=0}^{2^k-1} \sum_{j=0}^{M} (t-a) c \lambda_{ij} \psi_{ij}(t),$$

where $t \in [a, b]$ and

$$\psi_{nm} = \sqrt{2m+1} \frac{2^{k/2}}{\sqrt{b-a}} P_m \left(\frac{2^k(t-a)}{b-a} - n \right),$$

$$\frac{n(t-a)}{2^k} + a \leq t < \frac{(n+1)(t-a)}{2^k} + a,$$

with the shifted Legendre polynomials P_m.

The collocation points p_i, $1 \leq i \leq 2^k(M+1)$, are the roots of Chebyshev polynomials of degree $2^k(M+1)$. The resulting system of algebraic equations is solved to obtain the approximate solutions. Although the paper (Yousefi, Lotfi and Dehghan, 2011) discusses the general case when x and u are vector functions, for the sake of simplicity we outlined it here in one dimension.

A finite element method has been developed in (Agrawal, 2008b). The functional to be minimized has a special form of

$$J[x] = \int_a^b L\left(t, x(t), {}_aD_t^\alpha x(t)\right) dt$$

$$= \int_a^b \left[\frac{1}{2} A_1(t) \left({}_aD_t^\alpha x(t)\right)^2 + A_2(t) \left({}_aD_t^\alpha x(t)\right) x(t) \right.$$

$$\left. + \frac{1}{2} A_3(t) x^2(t) + A_4(t) {}_aD_t^\alpha x(t) + A_5(t) x(t) \right] dt.$$

The boundary conditions at both end-points are given. In this method, the time interval $[a, b]$ is divided into N equally spaced subintervals. Let $t_j = a + jh$ where $h = \frac{b-a}{N}$ and $j = 0, \ldots, N$. Then the functional is given by

$$J[x] = \sum_{j=1}^N \int_{t_{j-1}}^{t_j} L(t, x(t), {}_aD_t^\alpha x(t)) dt.$$

Now one can approximate x over subintervals by "shape" functions, e.g., splines, as

$$x(t) = N_j(t) x_{ej}, \qquad t \in [t_{j-1}, t_j], \ j = 1, \ldots, N,$$

and

$${}_aD_t^\alpha x(t) = N_j(t)({}_aD_t^\alpha x)_{ej}, \qquad t \in [t_{j-1}, t_j], \ j = 2, \ldots, N,$$

where N_j is the shape function at the corresponding subinterval, and x_{ej} and $({}_aD_t^\alpha x)_{ej}$ are the nodal values of the unknown function and its fractional derivatives. The fractional derivative at each point is also approximated using the Grünwald–Letnikov definition as an approximation, which is discussed in Chapter 8. The remaining process is straightforward.

Another work that is worth paying attention to is the use of a modified Grünwald–Letnikov approximation for left and right derivatives to discretize the Euler–Lagrange equation (Baleanu, Defterli and Agrawal, 2009). The approximations are carried out at the central points of a certain discretization of the time horizon. Namely, for $a = t_0 < t_1 < \ldots < t_n = b$,

$$_aD_t^\alpha x(t_{i-1/2}) \approx \frac{1}{h^\alpha} \sum_{k=0}^{i} (\omega_k^\alpha)\, x_{i-j}, \qquad i = 1, \ldots, n,$$

and

$$_tD_1^\alpha \lambda(t_{i+1/2}) \approx \frac{1}{h^\alpha} \sum_{k=0}^{n-i} (\omega_k^\alpha)\, \lambda_{i+j}, \qquad i = n-1, \ldots, 0,$$

where $(\omega_k^\alpha) = (-1)^k \binom{\alpha}{k} = \frac{\Gamma(k-\alpha)}{\Gamma(-\alpha)\Gamma(k+1)}$ and $x(t_{i-1/2}) = (x_{i-1} + x_i)/2$. Solving a system of $2n$ algebraic equations in $2n$ unknowns gives the approximate values of the unknown function on mesh points.

Numerical methods, nowadays, are easily implemented on computers that contain packages and tools to solve problems. Many problems in this book have been solved, e.g., in MATLAB®, using some predefined routines and solvers. The implemented methods are far from being an outstanding and a multipurpose solver. They have been designed for special problems and for a relevant problem they may need significant modifications. The only work, to the best of our knowledge, directed in the adaptation of the existing toolboxes is (Tricaud and Chen, 2010a). This work uses Oustaloup's approximation formula for fractional derivatives and transforms a fractional optimal control problem into a problem in which only derivatives of integer-order are present. Being a classical problem, it can be solved by **RIOTS-95**, a MATLAB® toolbox for optimal control problems.[1] The problem is to find a control that minimizes the functional

$$J[u] = G(x(a), x(b)) + \int_a^b L(t, x, u)dt$$

subject to the dynamic control system

$$_aD_t^\alpha x(t) = f(t, x, u),$$

and the initial condition $x(a) = x_a$. The control may be bounded,

$$u_{min} \leq u(t) \leq u_{max}.$$

---

[1]http://www.schwartz homo.com/RIOTS/

Also other constraints on the boundaries and/or state-control inequality constraints may be present. The idea – see (Tricaud and Chen, 2010a) for details – is to use a state-space approximation

$$_aD_t^\alpha x(t) \approx \begin{cases} \dot{z} = Az + Bu \\ x = Cz + Du, \end{cases}$$

and transform the problem to the minimization of

$$J[u] = G(Cz(a) + Du(a), Cz(b) + Du(b)) + \int_a^b L(t, Cz + Du, u)dt$$

such that

$$\dot{z}(t) = Az + B(f(t, Cz + Du, u)),$$

and the initial condition

$$z(a) = \frac{x_a \omega}{C\omega},$$

where $\omega = [1 \ \ 0 \ \cdots \ \ 0]^T$. The resulting setting is appropriate as an input for **RIOTS-95**.

Another approach to benefit the methods and tools of the classical theory has been introduced in (Jelicic and Petrovacki, 2009). The work is based on an approximation formula from (Atanacković and Stanković, 2008), which is improved and discussed in a very detailed way throughout our work. The control problem to be solved is the following:

$$J[u] = \int_0^1 L(t, x, u)dt \longrightarrow \min$$

subject to

$$\begin{cases} \dot{x}(t) + k\left(_aD_t^\alpha x(t)\right) = f(t, x, u) \\ x(0) = x_0. \end{cases}$$

Using the approximation

$$_aD_t^\alpha x(t) \approx At^{-\alpha} x(t) - \sum_{p=2}^N C(\alpha, p)t^{1-p-\alpha}V_p(t),$$

the problem is transformed into a classic integer-order problem,

$$J[u] = \int_0^1 L(t, x, u)dt \longrightarrow \min$$

subject to

$$\begin{cases} \dot{x}(t) + k\left(At^{-\alpha}x(t) - \sum_{p=2}^N C(\alpha, p)t^{1-p-\alpha}V_p(t)\right) = f(t, x, u) \\ \dot{V}_p(t) = (1-p)(t-a)^{p-2}x(t) \\ V_p(a) = 0, \qquad p = 2, \ldots, N \\ x(0) = x_0. \end{cases}$$

Chapter 6

Approximating fractional derivatives

This chapter is devoted to two approximations for the Riemann–Liouville, Caputo and Hadamard derivatives that are referred to as fractional operators afterwards. We introduce the expansions of fractional operators in terms of infinite sums involving only integer-order derivatives. These expansions are then used to approximate fractional operators in problems like fractional differential equations, fractional calculus of variations, fractional optimal control, etc. In this way, one can transform such problems into classical problems. Hereafter, a suitable method, that can be found in the classical literature, is employed to find an approximate solution for the original fractional problem. Here we focus mainly on the left derivatives and the details of extracting corresponding expansions. Right derivatives are given whenever it is needed to apply new techniques.

6.1 Riemann–Liouville derivative

6.1.1 *Approximation by a sum of integer-order derivatives*

Recall the definition of the left Riemann–Liouville derivative for $\alpha \in (0,1)$:

$$_aD_t^\alpha x(t) = \frac{1}{\Gamma(1-\alpha)} \frac{d}{dt} \int_a^t (t-\tau)^{-\alpha} x(\tau) d\tau. \tag{6.1}$$

The following theorem holds for any function x that is analytic in an interval $(c,d) \supset [a,b]$. See (Atanacković, Konjik and Pilipović, 2008) for a more detailed discussion and (Samko, Kilbas and Marichev, 1993), for a different proof.

Theorem 6.1. *Let (c,d), $-\infty < c < d < +\infty$, be an open interval in $\mathbb{R}$, and $[a,b] \subset (c,d)$ be such that for each $t \in [a,b]$ the closed ball $B_{b-a}(t)$,* *with center at t and radius $b - a$, lies in (c, d). If x is analytic in (c, d), then*

$$_aD_t^\alpha x(t) = \sum_{k=0}^{\infty} \frac{(-1)^{k-1}\alpha x^{(k)}(t)}{k!(k-\alpha)\Gamma(1-\alpha)}(t-a)^{k-\alpha}. \tag{6.2}$$

Proof.   Since $x(t)$ is analytic in (c, d) and $B_{b-a}(t) \subset (c, d)$ for any $\tau \in (a, t)$ with $t \in (a, b)$, the Taylor expansion of $x(\tau)$ at t is a convergent power series, i.e.,

$$x(\tau) = x(t - (t - \tau)) = \sum_{k=0}^{\infty} \frac{(-1)^k x^{(k)}(t)}{k!}(t-\tau)^k,$$

and then by (6.1)

$$_aD_t^\alpha x(t) = \frac{1}{\Gamma(1-\alpha)}\frac{d}{dt}\int_a^t \left((t-\tau)^{-\alpha}\sum_{k=0}^{\infty}\frac{(-1)^k x^{(k)}(t)}{k!}(t-\tau)^k\right)d\tau. \tag{6.3}$$

Since $(t - \tau)^{k-\alpha}x^{(k)}(t)$ is analytic, we can interchange integration with summation, so

$$_aD_t^\alpha x(t) = \frac{1}{\Gamma(1-\alpha)}\frac{d}{dt}\left(\sum_{k=0}^{\infty}\frac{(-1)^k x^{(k)}(t)}{k!}\int_a^t (t-\tau)^{k-\alpha}d\tau\right)$$

$$= \frac{1}{\Gamma(1-\alpha)}\frac{d}{dt}\sum_{k=0}^{\infty}\left(\frac{(-1)^k x^{(k)}(t)}{k!(k+1-\alpha)}(t-a)^{k+1-\alpha}\right)$$

$$= \sum_{k=0}^{\infty}\left(\frac{(-1)^k x^{(k+1)}(t)}{k!(k+1-\alpha)}\frac{(t-a)^{k+1-\alpha}}{\Gamma(1-\alpha)} + \frac{(-1)^k x^{(k)}(t)}{k!}\frac{(t-a)^{k-\alpha}}{\Gamma(1-\alpha)}\right)$$

$$= \frac{x(t)}{\Gamma(1-\alpha)}(t-a)^{-\alpha}$$

$$+ \frac{1}{\Gamma(1-\alpha)}\sum_{k=1}^{\infty}\left(\frac{(-1)^{k-1}}{(k-\alpha)(k-1)!} + \frac{(-1)^k}{k!}\right)x^{(k)}(t)(t-a)^{k-\alpha}.$$

Observe that

$$\frac{(-1)^{k-1}}{(k-\alpha)(k-1)!} + \frac{(-1)^k}{k!} = \frac{k(-1)^{k-1} + k(-1)^k - \alpha(-1)^k}{(k-\alpha)k!}$$

$$= \frac{(-1)^{k-1}\alpha}{(k-\alpha)k!},$$

since for any $k = 0, 1, 2, \ldots$ we have $k(-1)^{k-1} + k(-1)^k = 0$. Therefore, the expansion formula is reached as required.   □

For numerical purposes, a finite number of terms in (6.2) is used and one has

$$_aD_t^\alpha x(t) \approx \sum_{k=0}^{N} \frac{(-1)^{k-1}\alpha x^{(k)}(t)}{k!(k-\alpha)\Gamma(1-\alpha)}(t-a)^{k-\alpha}. \tag{6.4}$$

Remark 6.1. With the same assumptions of Theorem 6.1, we can expand $x(\tau)$ at t, where $\tau \in (t, b)$,

$$x(\tau) = x(t + (\tau - t)) = \sum_{k=0}^{\infty} \frac{x^{(k)}(t)}{k!}(\tau - t)^k,$$

and get the following approximation for the right Riemann–Liouville derivative:

$$_tD_b^\alpha x(t) \approx \sum_{k=0}^{N} \frac{-\alpha x^{(k)}(t)}{k!(k-\alpha)\Gamma(1-\alpha)}(b-t)^{k-\alpha}.$$

A proof for this expansion is available at (Samko, Kilbas and Marichev, 1993) that uses a similar relation for fractional integrals. The proof discussed here, however, allows us to extract an error term for this expansion easily.

6.1.2 *Approximation using moments of a function*

By "moments" of a function we have no physical or distributive senses in mind. The name comes from the fact that, during expansion, the terms of the form

$$V_p(t) := V_p(x(t)) = (1-p)\int_a^t (\tau - a)^{p-2}x(\tau)d\tau, \quad p \in \mathbb{N}, \tau \geq a, \tag{6.5}$$

appear to resemble the formulas of central moments (cf. (Atanacković and Stanković, 2008)). We assume that $V_p(x)$, $p \in \mathbb{N}$, denote the $(p-2)$th moment of a function $x \in AC^2[a, b]$.

The following lemma, that is given here without a proof, is the key relation to extract an expansion formula for Riemann–Liouville derivatives.

Lemma 6.1 (cf. Lemma 2.12 of (Diethelm, 2010)).
Let $x \in AC[a, b]$ and $0 < \alpha < 1$. Then the left Riemann–Liouville fractional derivative $_aD_t^\alpha x$ exists almost everywhere in $[a, b]$. Moreover, $_aD_t^\alpha x \in L_p[a, b]$ for $1 \leq p < \frac{1}{\alpha}$ and

$$_aD_t^\alpha x(t) = \frac{1}{\Gamma(1-\alpha)}\left[\frac{x(a)}{(t-a)^\alpha} + \int_a^t (t-\tau)^{-\alpha}\dot{x}(\tau)d\tau\right], \qquad t \in (a, b). \tag{6.6}$$

The same argument is valid for the right Riemann–Liouville derivative and

$$_tD_b^\alpha x(t) = \frac{1}{\Gamma(1-\alpha)}\left[\frac{x(b)}{(b-t)^\alpha} - \int_t^b (\tau - t)^{-\alpha}\dot{x}(\tau)d\tau\right], \qquad t \in (a,b).$$

Theorem 6.2 (cf. (Atanacković and Stanković, 2008)).
Let $x \in AC[a,b]$ and $0 < \alpha < 1$. Then the left Riemann–Liouville derivative can be expanded as

$$_aD_t^\alpha x(t) = A(\alpha)(t-a)^{-\alpha}x(t) + B(\alpha)(t-a)^{1-\alpha}\dot{x}(t)$$
$$- \sum_{p=2}^\infty C(\alpha,p)(t-a)^{1-p-\alpha}V_p(t), \tag{6.7}$$

where $V_p(t)$ is defined by (6.5) and

$$A(\alpha) = \frac{1}{\Gamma(1-\alpha)}\left(1 + \sum_{p=2}^\infty \frac{\Gamma(p-1+\alpha)}{\Gamma(\alpha)(p-1)!}\right),$$

$$B(\alpha) = \frac{1}{\Gamma(2-\alpha)}\left(1 + \sum_{p=1}^\infty \frac{\Gamma(p-1+\alpha)}{\Gamma(\alpha-1)p!}\right),$$

$$C(\alpha,p) = \frac{1}{\Gamma(2-\alpha)\Gamma(\alpha-1)}\frac{\Gamma(p-1+\alpha)}{(p-1)!}. \tag{6.8}$$

Remark 6.2. The proof of Theorem 6.2 is done by Atanacković and Stanković (2008) but, unfortunately, has a small mistake: the coefficient $A(\alpha)$, where we have an infinite sum, is not well defined since the series diverges. For a correct formulation and proof see our Theorem 6.3 and Remark 6.6.

The moments $V_p(t)$, $p = 2,3,\ldots$, are regarded as the solutions to the following system of differential equations:

$$\begin{cases} \dot{V}_p(t) = (1-p)(t-a)^{p-2}x(t) \\ V_p(a) = 0, \qquad p = 2,3,\ldots. \end{cases} \tag{6.9}$$

As before, a numerical approximation is achieved by taking only a finite number of terms in the series (6.7). We approximate the fractional derivative as

$$_aD_t^\alpha x(t) \approx A(t-a)^{-\alpha}x(t) + B(t-a)^{1-\alpha}\dot{x}(t) - \sum_{p=2}^N C(\alpha,p)(t-a)^{1-p-\alpha}V_p(t),$$
$$\tag{6.10}$$

where $A = A(\alpha, N)$ and $A = B(\alpha, N)$ are given by

$$A(\alpha, N) = \frac{1}{\Gamma(1 - \alpha)} \left(1 + \sum_{p=2}^{N} \frac{\Gamma(p - 1 + \alpha)}{\Gamma(\alpha)(p - 1)!} \right), \tag{6.11}$$

$$B(\alpha, N) = \frac{1}{\Gamma(2 - \alpha)} \left(1 + \sum_{p=1}^{N} \frac{\Gamma(p - 1 + \alpha)}{\Gamma(\alpha - 1)p!} \right). \tag{6.12}$$

Remark 6.3. The expansion (6.7) has been proposed in (Djordjevic and Atanacković, 2008) and an interesting, yet misleading, simplification has been made in (Atanacković and Stanković, 2008), which uses the fact that the infinite series $\sum_{p=1}^{\infty} \frac{\Gamma(p-1+\alpha)}{\Gamma(\alpha-1)p!}$ tends to -1 and concludes that $B(\alpha) = 0$ and thus

$$_aD_t^\alpha x(t) \approx A(\alpha, N)t^{-\alpha}x(t) - \sum_{p=2}^{N} C(\alpha, p)t^{1-p-\alpha}V_p(t). \tag{6.13}$$

In practice, however, we only use a finite number of terms in the series. Therefore

$$1 + \sum_{p=1}^{N} \frac{\Gamma(p - 1 + \alpha)}{\Gamma(\alpha - 1)p!} \neq 0,$$

and we keep here the approximation in the form of equation (6.10) (Pooseh, Almeida and Torres, 2013a). To be more precise, the values of $B(\alpha, N)$ for different choices of N and α are given in Table 6.1. It shows that even for a large N, when α tends to one, $B(\alpha, N)$ cannot be ignored. In Fig. 6.1, we plot $B(\alpha, N)$ as a function of N for different values of α.

Table 6.1 $B(\alpha, N)$ for different values of α and N.

N	4	7	15	30	70	120	170
$B(0.1, N)$	0.0310	0.0188	0.0095	0.0051	0.0024	0.0015	0.0011
$B(0.3, N)$	0.1357	0.0928	0.0549	0.0339	0.0188	0.0129	0.0101
$B(0.5, N)$	0.3085	0.2364	0.1630	0.1157	0.0760	0.0581	0.0488
$B(0.7, N)$	0.5519	0.4717	0.3783	0.3083	0.2396	0.2040	0.1838
$B(0.9, N)$	0.8470	0.8046	0.7481	0.6990	0.6428	0.6092	0.5884
$B(0.99, N)$	0.9849	0.9799	0.9728	0.9662	0.9582	0.9531	0.9498

Remark 6.4. Similar computations give rise to an expansion formula for $_tD_b^\alpha$, the right Riemann–Liouville fractional derivative:

$$_tD_b^\alpha x(t) \approx A(b-t)^{-\alpha}x(t) - B(b-t)^{1-\alpha}\dot{x}(t) - \sum_{p=2}^{N} C(\alpha, p)(b-t)^{1-p-\alpha}W_p(t), \tag{6.14}$$

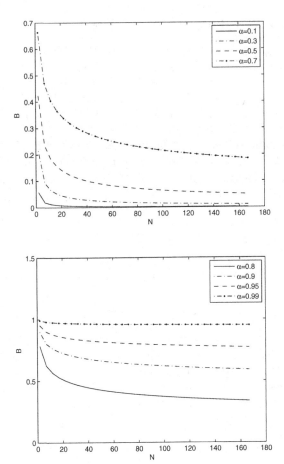

Fig. 6.1 $B(\alpha, N)$ for different values of α and N.

where

$$W_p(t) = (1-p) \int_t^b (b-\tau)^{p-2} x(\tau) d\tau.$$

The coefficients $A = A(\alpha, N)$ and $B = B(\alpha, N)$ are the same as (6.11) and (6.12), respectively, and $C(\alpha, p)$ is given by (6.8).

Remark 6.5. As stated before, Caputo derivatives are closely related to those of Riemann–Liouville. For any function, x, and for $\alpha \in (0, 1)$, if these two kind of fractional derivatives exist, then we have

$$\,_{a}^{C}D_{t}^{\alpha}x(t) = \,_{a}D_{t}^{\alpha}x(t) - \frac{x(a)}{\Gamma(1-\alpha)}(t-a)^{-\alpha}$$

and

$$\,_{t}^{C}D_{b}^{\alpha}x(t) = \,_{t}D_{b}^{\alpha}x(t) - \frac{x(b)}{\Gamma(1-\alpha)}(b-t)^{-\alpha}.$$

Using these relations, we can easily construct approximation formulas for left and right Caputo fractional derivatives:

$$\,_{a}^{C}D_{t}^{\alpha}x(t) \approx A(\alpha,N)(t-a)^{-\alpha}x(t) + B(\alpha,N)(t-a)^{1-\alpha}\dot{x}(t)$$
$$- \sum_{p=2}^{N}C(\alpha,p)(t-a)^{1-p-\alpha}V_{p}(t) - \frac{x(a)}{\Gamma(1-\alpha)}(t-a)^{-\alpha}.$$

Formula (6.7) consists of two parts: an infinite series and two terms including the first derivative and the function itself. It can be generalized to contain derivatives of higher-order.

Theorem 6.3. *Fix $n \in \mathbb{N}$ and let $x \in C^{n}[a,b]$. Then,*

$$\,_{a}D_{t}^{\alpha}x(t) = \frac{1}{\Gamma(1-\alpha)}(t-a)^{-\alpha}x(t) + \sum_{i=1}^{n-1}A(\alpha,i)(t-a)^{i-\alpha}x^{(i)}(t)$$
$$+ \sum_{p=n}^{\infty}\left[\frac{-\Gamma(p-n+1+\alpha)}{\Gamma(-\alpha)\Gamma(1+\alpha)(p-n+1)!}(t-a)^{-\alpha}x(t)\right.$$
$$\left.+ B(\alpha,p)(t-a)^{n-1-p-\alpha}V_{p}(t)\right], \quad (6.15)$$

where

$$A(\alpha,i) = \frac{1}{\Gamma(i+1-\alpha)}\left[1 + \sum_{p=n-i}^{\infty}\frac{\Gamma(p-n+1+\alpha)}{\Gamma(\alpha-i)(p-n+i+1)!}\right],$$

$i = 1,\ldots,n-1,$

$$B(\alpha,p) = \frac{\Gamma(p-n+1+\alpha)}{\Gamma(-\alpha)\Gamma(1+\alpha)(p-n+1)!},$$
$$V_{p}(t) = (p-n+1)\int_{a}^{t}(\tau-a)^{p-n}x(\tau)d\tau.$$

Proof.   Successive integrating by parts in (6.6) gives

$$_aD_t^\alpha x(t) = \frac{x(a)}{\Gamma(1-\alpha)}(t-a)^{-\alpha} + \frac{\dot{x}(a)}{\Gamma(2-\alpha)}(t-a)^{1-\alpha}$$

$$+ \cdots + \frac{x^{(n-1)}(a)}{\Gamma(n-\alpha)}(t-a)^{n-1-\alpha}$$

$$+ \frac{1}{\Gamma(1-\alpha)} \int_a^t (t-\tau)^{n-1-\alpha} x^{(n)}(\tau)d\tau.$$

Using the binomial theorem, we expand the integral term as

$$\int_a^t (t-\tau)^{n-1-\alpha} x^{(n)}(\tau)d\tau$$

$$= (t-a)^{n-1-\alpha} \sum_{p=0}^\infty \frac{\Gamma(p-n+1+\alpha)}{\Gamma(1-n+\alpha)p!(t-a)^p} \int_a^t (\tau-a)^p x^{(n)}(\tau)d\tau.$$

Splitting the sum into $p = 0$ and $p = 1 \ldots \infty$, and integrating by parts the last integral, we get

$$_aD_t^\alpha x(t) = \frac{(t-a)^{-\alpha}}{\Gamma(1-\alpha)} x(a) + \cdots + \frac{(t-a)^{n-2-\alpha}}{\Gamma(n-1-\alpha)} x^{(n-2)}(a)$$

$$+ \frac{(t-a)^{n-1-\alpha}}{\Gamma(n-\alpha)} x^{(n-2)}(t) \left[1 + \sum_{p=1}^\infty \frac{\Gamma(p-n+1+\alpha)}{\Gamma(-n+1+\alpha)p!} \right]$$

$$+ \frac{(t-a)^{n-1-\alpha}}{\Gamma(n-1-\alpha)}$$

$$\times \sum_{p=1}^\infty \frac{\Gamma(p-n+1+\alpha)}{\Gamma(-n+2+\alpha)(p-1)!(t-a)^p} \int_a^t (\tau-a)^{p-1} x^{(n-1)}(\tau)d\tau.$$

The rest of the proof follows a similar routine, i.e., by splitting the sum into two parts, the first term and the rest, and integrating by parts the last integral until x appears in the integrand.   □

Remark 6.6. The series that appear in $A(\alpha, i)$ is convergent for all $i \in \{1, \ldots, n-1\}$. Fix an i and observe that

$$\sum_{p=n-i}^\infty \frac{\Gamma(p-n+1+\alpha)}{\Gamma(\alpha-i)(p-n+i+1)!} = \sum_{p=1}^\infty \frac{\Gamma(p+\alpha-i)}{\Gamma(\alpha-i)p!} = {}_1F_0(\alpha-i,1) - 1,$$

where ${}_1F_0$ stands for a hypergeometric function (Andrews, Askey and Roy, 1999). Since $i > \alpha$, ${}_1F_0(\alpha-i,1)$ converges by Theorem 2.1.1 of (Andrews, Askey and Roy, 1999).

In practice we only use finite sums and for $A(\alpha, i)$ we can easily compute the truncation error. Although this is a partial error, it gives a good intuition of why this approximation works well. Using the fact that $_1F_0(a, 1) = 0$ if $a < 0$ (cf. Eq. (2.1.6) in (Andrews, Askey and Roy, 1999)), we have

$$
\frac{1}{\Gamma(i+1-\alpha)} \sum_{p=N+1}^{\infty} \frac{\Gamma(p-n+1+\alpha)}{\Gamma(\alpha-i)(p-n+i+1)!}
$$

$$
= \frac{1}{\Gamma(i+1-\alpha)} \left({}_1F_0(\alpha-i, 1) - \sum_{p=0}^{N-n+i+1} \frac{\Gamma(p+\alpha-i)}{\Gamma(\alpha-i)p!} \right)
$$

$$
= \frac{-1}{\Gamma(i+1-\alpha)} \sum_{p=0}^{N-n+i+1} \frac{\Gamma(p+\alpha-i)}{\Gamma(\alpha-i)p!}.
$$

$$(6.16)$$

In Table 6.2 we give some values for this error, with $\alpha = 0.5$ and different values for i and $N - n$.

Table 6.2   The truncation error (6.16) of $A(\alpha, i)$ for $\alpha = 0.5$, that is, $A(\alpha, i) - A(\alpha, i, N)$ with $A(\alpha, i, N)$ given by (6.18).

i $\diagdown$ $N-n$	0	5	10	15	20
1	-0.4231	-0.2364	-0.1819	-0.1533	-0.1350
2	0.04702	0.009849	0.004663	0.002838	0.001956
3	-0.007052	-0.0006566	-0.0001999	-0.00008963	-0.00004890
4	0.001007	0.00004690	0.000009517	0.000003201	0.000001397

Remark 6.7. Using Euler's reflection formula, one can define $B(\alpha, p)$ of Theorem 6.3 as

$$
B(\alpha, p) = \frac{-\sin(\pi\alpha)\Gamma(p-n+1+\alpha)}{\pi(p-n+1)!}.
$$

For numerical purposes, only finite sums are taken to approximate fractional derivatives. Therefore, for a fixed $n \in \mathbb{N}$ and $N \geq n$, one has

$$
{}_aD_t^{\alpha}x(t) \approx \sum_{i=0}^{n-1} A(\alpha, i, N)(t-a)^{i-\alpha}x^{(i)}(t) + \sum_{p=n}^{N} B(\alpha, p)(t-a)^{n-1-p-\alpha}V_p(t),
$$

$$(6.17)$$

where

$$
A(\alpha, i, N) = \frac{1}{\Gamma(i+1-\alpha)} \left[1 + \sum_{p=2}^{N} \frac{\Gamma(p-n+1+\alpha)}{\Gamma(\alpha-i)(p-n+i+1)!} \right], \quad (6.18)
$$

$i = 0, \ldots, n-1,$

$$B(\alpha, p) = \frac{\Gamma(p - n + 1 + \alpha)}{\Gamma(-\alpha)\Gamma(1 + \alpha)(p - n + 1)!},$$

$$V_p(t) = (p - n + 1) \int_a^t (\tau - a)^{p-n} x(\tau) d\tau.$$

Similarly, we can deduce an expansion formula for the right fractional derivative.

Theorem 6.4. *Fix $n \in \mathbb{N}$ and $x \in C^n[a, b]$. Then,*

$$_tD_b^\alpha x(t) = \frac{1}{\Gamma(1 - \alpha)}(b - t)^{-\alpha} x(t) + \sum_{i=1}^{n-1} A(\alpha, i)(b - t)^{i-\alpha} x^{(i)}(t)$$

$$+ \sum_{p=n}^{\infty} \left[\frac{-\Gamma(p - n + 1 + \alpha)}{\Gamma(-\alpha)\Gamma(1 + \alpha)(p - n + 1)!}(b - t)^{-\alpha} x(t) \right.$$

$$\left. + B(\alpha, p)(b - t)^{n-1-\alpha-p} W_p(t) \right],$$

where

$$A(\alpha, i) = \frac{(-1)^i}{\Gamma(i + 1 - \alpha)} \left[1 + \sum_{p=n-i}^{\infty} \frac{\Gamma(p - n + 1 + \alpha)}{\Gamma(-i + \alpha)(p - n + 1 + i)!} \right],$$

$i = 1, \ldots, n-1,$

$$B(\alpha, p) = \frac{(-1)^n \Gamma(p - n + 1 + \alpha)}{\Gamma(-\alpha)\Gamma(1 + \alpha)(p - n + 1)!},$$

$$W_p(t) = (p - n + 1) \int_t^b (b - \tau)^{p-n} x(\tau) d\tau.$$

Proof.   Analogous to the proof of Theorem 6.3.   □

6.1.3   *Numerical evaluation of fractional derivatives*

In (Podlubny, 1999) a numerical method to evaluate fractional derivatives is given based on the Grünwald–Letnikov definition of fractional derivatives. It uses the fact that for a large class of functions, the Riemann–Liouville and the Grünwald–Letnikov definitions are equivalent. We claim that the approximations discussed so far provide a good tool to compute numerically the fractional derivatives of given functions. For functions whose higher-order derivatives are easily available, we can freely choose between approximations (6.4) or (6.10). But in the case that difficulties arise in computing higher-order derivatives, we choose the approximation (6.10) that needs only the values of the first derivative and function itself. Even if the first derivative is not easily computable, we can use the approximation given by (6.13) with large values for N and α not so close to one. As an example, we compute $_aD_t^{\alpha}x(t)$, with $\alpha = \frac{1}{2}$, for $x(t) = t^4$ and $x(t) = e^{2t}$. The exact formulas of the derivatives are derived from

$$_0D_t^{0.5}(t^n) = \frac{\Gamma(n+1)}{\Gamma(n+1-0.5)}t^{n-0.5} \quad \text{and} \quad _0D_t^{0.5}(e^{\lambda t}) = t^{-0.5}E_{1,1-0.5}(\lambda t),$$

where $E_{\alpha,\beta}$ is the two parameter Mittag–Leffler function (Podlubny, 1999). Figure 6.2 shows the results using approximation (6.4) with error E computed by (1.3). As we can see, the third approximations are reasonably accurate for both cases. Indeed, for $x(t) = t^4$, the approximation with $N = 4$ coincides with the exact solution because the derivatives of order five and more vanish. The same computations are carried out using approximation (6.10). In this case, given a function x, we can compute V_p by definition or integrate the system (6.9) analytically or by any numerical integrator. As it is clear from Fig. 6.3, one can get better results by using larger values of N. Comparing Figs. 6.2 and 6.3, we find out that the approximation (6.4) shows a faster convergence. Observe that both functions are analytic and it is easy to compute higher-order derivatives. The approximation (6.4) fails for non-analytic functions as stated in (Atanacković and Stanković, 2008).

Remark 6.8. A closer look at (6.4) and (6.10) reveals that in both cases the approximations are not computable at a and b for the left and right fractional derivatives, respectively. At these points we assume that it is possible to extend them continuously to the closed interval $[a, b]$.

In what follows, we show that by omitting the first derivative from the expansion, as done in (Atanacković and Stanković, 2008), one may lose a considerable accuracy in computation. Once again, we compute the fractional derivatives of $x(t) = t^4$ and $x(t) = e^{2t}$, but this time we use the approximation given by (6.13). Figure 6.4 summarizes the results. The expansion up to the first derivative gives a more realistic approximation using a quite small N, 3 in this case. To show how the appearance of higher-order derivatives in generalization (6.15) gives better results, we evaluate fractional derivatives of $x(t) = t^4$ and $x(t) = e^{2t}$ for different values of n. We consider $n = 1, 2, 3$, $N = 6$ for $x(t) = t^4$ (Fig. 6.5(a)) and $N = 4$ for $x(t) = e^{2t}$ (Fig. 6.5(b)).

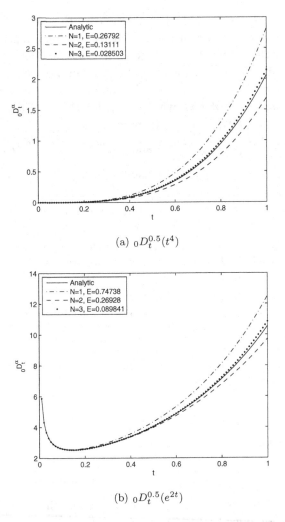

(a) $_0D_t^{0.5}(t^4)$

(b) $_0D_t^{0.5}(e^{2t})$

Fig. 6.2   Analytic (solid line) *versus* numerical approximation (6.4).

6.1.4   *Fractional derivatives of tabular data*

In many situations, the function itself is not accessible in a closed form, but as a tabular data for discrete values of the independent variable. Thus, we cannot use the definition to compute the fractional derivative directly. Approximation (6.10) that uses the function and its first derivative to

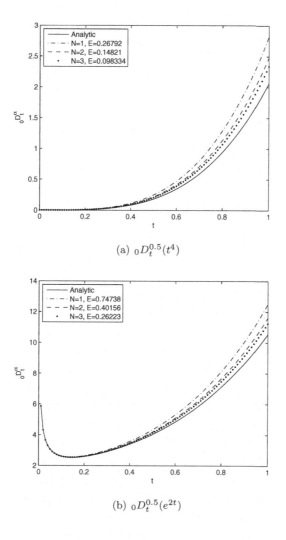

(a) $_0D_t^{0.5}(t^4)$

(b) $_0D_t^{0.5}(e^{2t})$

Fig. 6.3 Analytic (solid line) *versus* numerical approximation (6.10).

evaluate the fractional derivative, seems to be a good candidate in those cases. Suppose that we know the values of $x(t_i)$ on $n + 1$ distinct points in a given interval $[a, b]$, i.e., for t_i, $i = 0, 1, \ldots, n$, with $t_0 = a$ and $t_n = b$. According to formula (6.10), the value of the fractional derivative of x at

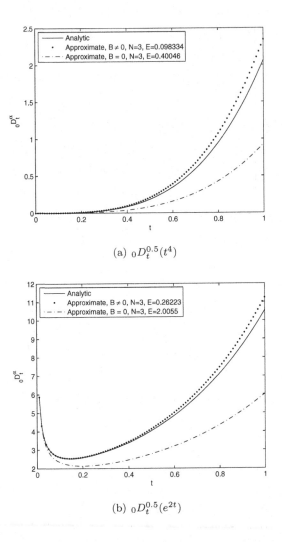

(a) $_0D_t^{0.5}(t^4)$

(b) $_0D_t^{0.5}(e^{2t})$

Fig. 6.4   Comparison of approximation (6.10) proposed here and approximation (6.13) of (Atanacković and Stanković, 2008).

each point t_i is given approximately by

$$_aD_{t_i}^\alpha x(t_i) \approx A(\alpha, N)(t_i - a)^{-\alpha}x(t_i) + B(\alpha, N)(t_i - a)^{1-\alpha}\dot{x}(t_i)$$
$$- \sum_{p=2}^{N} C(p, \alpha)(t_i - a)^{1-p-\alpha}V_p(t_i).$$

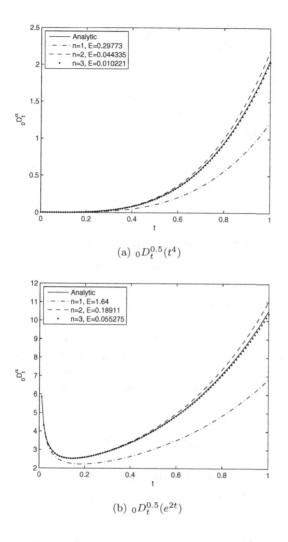

(a) $_0D_t^{0.5}(t^4)$

(b) $_0D_t^{0.5}(e^{2t})$

Fig. 6.5 Analytic (solid line) *versus* numerical approximation (6.15).

The values of $x(t_i)$, $i = 0, 1, \ldots, n$, are given. A good approximation for $\dot{x}(t_i)$ can be obtained using the forward, centered, or backward difference approximation of the first-order derivative (Stoer and Bulirsch, 2002). For $V_p(t_i)$ one can either use the definition and compute the integral numerically, i.e., $V_p(t_i) = \int_a^{t_i} (1-p)(\tau - a)^{p-2} x(\tau) d\tau$, or it is possible to solve (6.9) as an initial value problem. All required computations are straightforward and only need to be implemented with the desired accuracy. The only thing to be careful with is choosing a good order, N, in the formula (6.10). Because no value of N, guaranteeing the error to be smaller than a certain preassigned number, is known *a priori*, we start with some prescribed value for N and increase it step by step. In each step we compare, using an appropriate norm, the result with the one of the previous step. For instance, one can use the Euclidean norm $\|({}_aD_t^\alpha)^{new} - ({}_aD_t^\alpha)^{old}\|_2$ and terminate the procedure when its value is smaller than a predefined ϵ. For illustrative purposes, we compute the fractional derivatives of order $\alpha = 0.5$ for tabular data extracted from $x(t) = t^4$ and $x(t) = e^{2t}$. The results are given in Fig. 6.6.

6.1.5 *Applications to fractional differential equations*

The classical theory of ordinary differential equations is a well developed field with many tools available for numerical purposes. Using the approximations (6.4) and (6.10), one can transform a fractional ordinary differential equation into a classical ODE (ordinary differential equation).

We should mention here that, using (6.4), derivatives of higher-order appear in the resulting ODE, while we only have a limited number of initial or boundary conditions available. In this case the value of N, the order of approximation, should be equal to the number of given conditions. If we choose a larger N, we will encounter a lack of initial or boundary conditions. This problem is not present in the case in which we use the approximation (6.10), because the initial values for the auxiliary variables V_p, $p = 2, 3, \ldots$, are known and we do not require any extra information.

Consider, as an example, the following initial value problem:

$$\begin{cases} {}_0D_t^{0.5}x(t) + x(t) = t^2 + \frac{2}{\Gamma(2.5)}t^{\frac{3}{2}}, \\ x(0) = 0. \end{cases} \tag{6.19}$$

We know that ${}_0D_t^{0.5}(t^2) = \frac{2}{\Gamma(2.5)}t^{\frac{3}{2}}$. Therefore, the analytic solution for system (6.19) is $x(t) = t^2$. Since only one initial condition is available, we can only expand the fractional derivative up to the first derivative in (6.4). One has

$$\begin{cases} 1.5642\, t^{-0.5}x(t) + 0.5642\, t^{0.5}\dot{x}(t) = t^2 + 1.5045\, t^{1.5}, \\ x(0) = 0. \end{cases} \tag{6.20}$$

This is a classical initial value problem and can be easily treated numerically. The solution is drawn in Fig. 6.7(a). As expected, the result is not satisfactory. Let us now use the approximation given by (6.10).

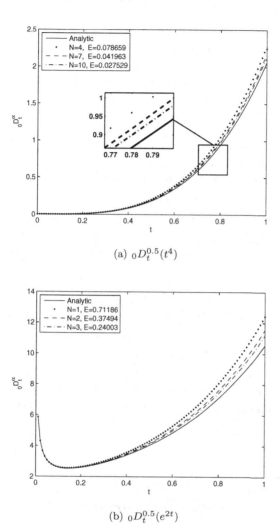

(a) $_0D_t^{0.5}(t^4)$

(b) $_0D_t^{0.5}(e^{2t})$

Fig. 6.6   Fractional derivatives of tabular data.

The system in (6.19) becomes

$$
\begin{cases}
A(N)t^{-0.5}x(t) + B(N)t^{0.5}\dot{x}(t) - \sum_{p=2}^{N} C(p)t^{0.5-p}V_p + x(t) \\
\quad = t^2 + \frac{2}{\Gamma(2.5)}t^{1.5}, \\
\dot{V}_p(t) = (1-p)(t-a)^{p-2}x(t), \quad p = 2, 3, \ldots, N, \\
x(0) = 0, \\
V_p(0) = 0, \quad p = 2, 3, \ldots, N.
\end{cases}
\tag{6.21}
$$

We solve this initial value problem for $N = 7$. The MATLAB® ode45 built-in function is used to integrate system (6.21). The solution is given in Fig. 6.7(b) and shows a better approximation when compared with (6.20).

Remark 6.9. To show the difference caused by the appearance of the first derivative in formula (6.10), we solve the initial value problem (6.19) with $B(\alpha, N) = 0$. Since the original fractional differential equation does not depend on integer-order derivatives of function x, i.e., it has the form

$$
{}_aD_t^\alpha x(t) + f(x, t) = 0,
$$

by (6.13) the dependence to derivatives of x vanishes. In this case, one needs to apply the operator ${}_aD_t^{1-\alpha}$ to the above equation and obtain

$$
\dot{x}(t) + {}_aD_t^{1-\alpha}[f(x, t)] = 0.
$$

Nevertheless, we can use (6.10) directly without any trouble. Figure 6.8 shows that at least for a moderate accurate method, like the MATLAB® routine ode45, taking $B(\alpha, N) \neq 0$ into account gives a better approximation.

6.2   Hadamard derivatives

For Hadamard derivatives, the expansions can be obtained in quite a similar way and are introduced next (Pooseh, Almeida and Torres, 2012c).

6.2.1   *Approximation by a sum of integer-order derivatives*

Assume that function x admits derivatives of any order. Then expansion formulas for the Hadamard fractional integrals and derivatives of x, in terms of its integer-order derivatives, are given in (Butzer, Kilbas and Trujillo, 2003, Theorem 17):

$$
{}_0\mathcal{I}_t^\alpha x(t) = \sum_{k=0}^{\infty} S(-\alpha, k)t^k x^{(k)}(t)
$$

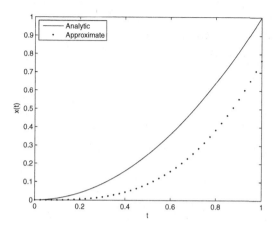

(a) Exact *versus* Approximation (6.4).

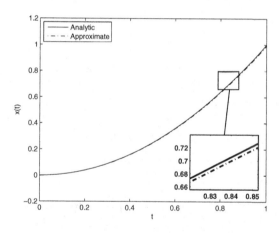

(b) Exact *versus* Approximation (6.10).

Fig. 6.7 Two approximations applied to fractional differential equation (6.19).

and

$$_0\mathcal{D}_t^\alpha x(t) = \sum_{k=0}^{\infty} S(\alpha, k) t^k x^{(k)}(t),$$

where

$$S(\alpha, k) = \frac{1}{k!} \sum_{j=1}^{k} (-1)^{k-j} \binom{k}{j} j^\alpha$$

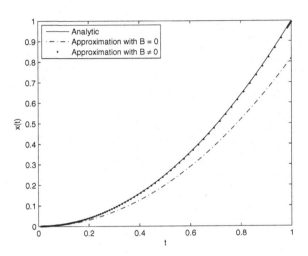

Fig. 6.8   Comparison of our approach to that of (Atanacković and Stanković, 2008).

is the Stirling function.

As approximations we truncate infinite sums at an appropriate order N and get the following formulas:

$$_0\mathcal{I}_t^\alpha x(t) \approx \sum_{k=0}^{N} S(-\alpha, k) t^k x^{(k)}(t)$$

and

$$_0\mathcal{D}_t^\alpha x(t) \approx \sum_{k=0}^{N} S(\alpha, k) t^k x^{(k)}(t).$$

6.2.2   *Approximation using moments of a function*

The same idea of expanding Riemann–Liouville derivatives, with slightly different techniques, is used to derive expansion formulas for left and right Hadamard derivatives. The following lemma is the basis for such new relations.

Lemma 6.2. *Let $\alpha \in (0, 1)$ and x be an absolutely continuous function on $[a, b]$. Then the Hadamard fractional derivatives may be expressed by*

$$_a\mathcal{D}_t^\alpha x(t) = \frac{x(a)}{\Gamma(1 - \alpha)} \left(\ln \frac{t}{a} \right)^{-\alpha} + \frac{1}{\Gamma(1 - \alpha)} \int_a^t \left(\ln \frac{t}{\tau} \right)^{-\alpha} \dot{x}(\tau) d\tau \quad (6.22)$$

and

$$_t\mathcal{D}_b^\alpha x(t) = \frac{x(b)}{\Gamma(1-\alpha)}\left(\ln\frac{b}{t}\right)^{-\alpha} - \frac{1}{\Gamma(1-\alpha)}\int_t^b \left(\ln\frac{\tau}{t}\right)^{-\alpha}\dot{x}(\tau)d\tau. \quad (6.23)$$

A proof of this lemma for an arbitrary $\alpha > 0$ can be found in (Kilbas, 2001, Theorem 3.2).

Applying similar techniques as presented in Theorem 6.3 to the formulas (6.22) and (6.23) gives the following theorem.

Theorem 6.5. *Let* $n \in \mathbb{N}$, $0 < a < b$ *and* $x : [a, b] \to \mathbb{R}$ *be a function of class* C^{n+1}. *Then*

$$_a\mathcal{D}_t^\alpha x(t) \approx \sum_{i=0}^n A_i(\alpha, N)\left(\ln\frac{t}{a}\right)^{i-\alpha} x_{i,0}(t)$$

$$+ \sum_{p=n+1}^N B(\alpha, p)\left(\ln\frac{t}{a}\right)^{n-\alpha-p} V_p(t)$$

with

$$A_i(\alpha, N) = \frac{1}{\Gamma(i+1-\alpha)}\left[1 + \sum_{p=n-i+1}^N \frac{\Gamma(p+\alpha-n)}{\Gamma(\alpha-i)(p-n+i)!}\right],$$

$i \in \{0, \ldots, n\}$,

$$B(\alpha, p) = \frac{\Gamma(p+\alpha-n)}{\Gamma(-\alpha)\Gamma(1+\alpha)(p-n)!}, \quad p \in \{n+1, \ldots\},$$

$$V_p(t) = \int_a^t (p-n)\left(\ln\frac{\tau}{a}\right)^{p-n-1}\frac{x(\tau)}{\tau}d\tau, \quad p \in \{n+1, \ldots\}.$$

Remark 6.10. The right Hadamard fractional derivative can be expanded in the same way. This gives the following approximation:

$$_t\mathcal{D}_b^\alpha x(t) \approx A(\alpha, N)\left(\ln\frac{b}{t}\right)^{-\alpha} x(t) - B(\alpha, N)\left(\ln\frac{b}{t}\right)^{1-\alpha} t\dot{x}(t)$$

$$- \sum_{p=2}^N C(\alpha, p)\left(\ln\frac{b}{t}\right)^{1-\alpha-p} W_p(t),$$

with

$$W_p(t) = (1-p)\int_t^b \left(\ln\frac{b}{\tau}\right)^{p-2}\frac{x(\tau)}{\tau}d\tau.$$

Remark 6.11. In the particular case $n = 1$, one obtains from Theorem 6.5 that

$$_a\mathcal{D}_t^\alpha x(t) \approx A(\alpha, N) \left(\ln \frac{t}{a} \right)^{-\alpha} x(t) + B(\alpha, N) \left(\ln \frac{t}{a} \right)^{1-\alpha} t\dot{x}(t)$$

$$+ \sum_{p=2}^N C(\alpha, p) \left(\ln \frac{t}{a} \right)^{1-\alpha-p} V_p(t) \tag{6.24}$$

with

$$A(\alpha, N) = \frac{1}{\Gamma(1-\alpha)} \left(1 + \sum_{p=2}^N \frac{\Gamma(p+\alpha-1)}{\Gamma(\alpha)(p-1)!} \right),$$

$$B(\alpha, N) = \frac{1}{\Gamma(2-\alpha)} \left(1 + \sum_{p=1}^N \frac{\Gamma(p+\alpha-1)}{\Gamma(\alpha-1)p!} \right).$$

6.2.3 *Examples*

In this section we apply (6.24) to compute fractional derivatives, of order $\alpha = 0.5$, for $x(t) = \ln(t)$ and $x(t) = t^4$. The exact Hadamard fractional derivative is available for $x(t) = \ln(t)$ and we have

$$_1\mathcal{D}_t^{0.5}(\ln(t)) = \frac{\sqrt{\ln t}}{\Gamma(1.5)}.$$

For $x(t) = t^4$ only an approximation of Hadamard fractional derivative is found in the literature:

$$_1\mathcal{D}_t^{0.5} t^4 \approx \frac{1}{\Gamma(0.5)\sqrt{\ln t}} + \frac{0.5908179503}{\Gamma(0.5)} 4t^4 \text{erf}(3\sqrt{\ln t}),$$

where erf in the so-called Gauss error function,

$$\text{erf}(t) = \frac{1}{\sqrt{\pi}} \int_0^t e^{-\tau^2} d\tau.$$

The results of applying (6.24) to evaluate fractional derivatives are depicted in Fig. 6.9.

As another example, we consider the following fractional differential equation involving a Hadamard fractional derivative:

$$\begin{cases} _1\mathcal{D}_t^{0.5} x(t) + x(t) = \dfrac{\sqrt{x(t)}}{\Gamma(1.5)} + \ln t \\ x(1) = 0. \end{cases} \tag{6.25}$$

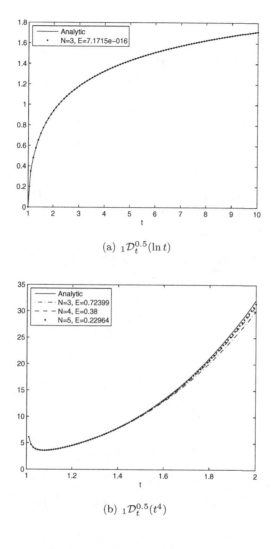

(a) $_1\mathcal{D}_t^{0.5}(\ln t)$

(b) $_1\mathcal{D}_t^{0.5}(t^4)$

Fig. 6.9   Analytic *versus* numerical approximation (6.24).

Obviously, $x(t) = \ln t$ is a solution for (6.25). Since we have only one initial condition, we replace the operator $_1\mathcal{D}_t^{0.5}$ by the expansion with $n = 1$ and thus obtaining

$$
\begin{cases}
\left[1 + A_0(\ln t)^{-0.5}\right]x(t) + A_1(\ln t)^{0.5}t\dot{x}(t) + \sum_{p=2}^{N} B(0.5,p)(\ln t)^{0.5-p}V_p(t) \\
\quad = \frac{\sqrt{x(t)}}{\Gamma(1.5)} + \ln t, \\
\dot{V}_p(t) = (p-1)(\ln t)^{p-2}\dfrac{x(t)}{t}, \quad p = 2,3,\ldots,N, \\
x(1) = 0, \\
V_p(1) = 0, \quad p = 2,3,\ldots,N.
\end{cases}
$$

$$(6.26)$$

In Fig. 6.10 we compare the analytical solution of problem (6.25) with the numerical result for $N = 2$ in (6.26).

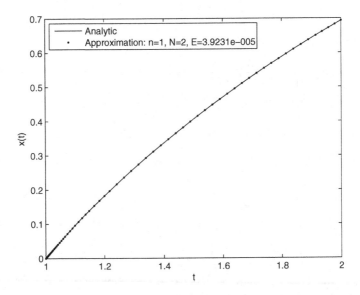

Fig. 6.10   Analytic *versus* numerical approximation for problem (6.25) with one initial condition.

6.3   Error analysis

When we approximate an infinite series by a finite sum, the choice of the order of approximation is a key question. Having an estimate knowledge of truncation errors, one can choose properly up to which order the approximations should be made to suit the accuracy requirements. In this section we study the errors of the approximations presented so far.

Separation of an error term in (6.3) ends in

$$
{}_aD_t^\alpha x(t) = \frac{1}{\Gamma(1-\alpha)} \frac{d}{dt} \int_a^t \left((t-\tau)^{-\alpha} \sum_{k=0}^N \frac{(-1)^k x^{(k)}(t)}{k!} (t-\tau)^k \right) d\tau
$$
$$
+ \frac{1}{\Gamma(1-\alpha)} \frac{d}{dt} \int_a^t \left((t-\tau)^{-\alpha} \sum_{k=N+1}^\infty \frac{(-1)^k x^{(k)}(t)}{k!} (t-\tau)^k \right) d\tau. \quad (6.27)
$$

The first term in (6.27) gives (6.4) directly and the second term is the error caused by truncation. The next step is to give a local upper bound for this error, $E_{tr}(t)$.

The series

$$
\sum_{k=N+1}^\infty \frac{(-1)^k x^{(k)}(t)}{k!} (t-\tau)^k, \quad \tau \in (a,t), \quad t \in (a,b),
$$

is the remainder of the Taylor expansion of $x(\tau)$ and thus bounded by $\left| \frac{M}{(N+1)!} (t-\tau)^{N+1} \right|$ in which

$$
M = \max_{\tau \in [a,t]} |x^{(N+1)}(\tau)|.
$$

Then,

$$
E_{tr}(t) \leq \left| \frac{M}{\Gamma(1-\alpha)(N+1)!} \frac{d}{dt} \int_a^t (t-\tau)^{N+1-\alpha} d\tau \right|
$$
$$
= \frac{M}{\Gamma(1-\alpha)(N+1)!} (t-a)^{N+1-\alpha}.
$$

In order to estimate a truncation error for approximation (6.10), the expansion procedure is carried out with separation of N terms in binomial expansion as

$$
\left(1 - \frac{\tau-a}{t-a} \right)^{1-\alpha} = \sum_{p=0}^\infty \frac{\Gamma(p-1+\alpha)}{\Gamma(\alpha-1)p!} \left(\frac{\tau-a}{t-a} \right)^p
$$
$$
= \sum_{p=0}^N \frac{\Gamma(p-1+\alpha)}{\Gamma(\alpha-1)p!} \left(\frac{\tau-a}{t-a} \right)^p + R_N(\tau), \quad (6.28)
$$

where

$$
R_N(\tau) = \sum_{p=N+1}^\infty \frac{\Gamma(p-1+\alpha)}{\Gamma(\alpha-1)p!} \left(\frac{\tau-a}{t-a} \right)^p.
$$

Integration by parts on the right-hand side of (6.6) gives

$$_aD_t^\alpha x(t) = \frac{x(a)}{\Gamma(1-\alpha)}(t-a)^{-\alpha} + \frac{\dot{x}(a)}{\Gamma(2-\alpha)}(t-a)^{1-\alpha}$$
$$+ \frac{1}{\Gamma(2-\alpha)}\int_a^t (t-\tau)^{1-\alpha}\ddot{x}(\tau)d\tau. \tag{6.29}$$

Substituting (6.28) into (6.29), we get

$$_aD_t^\alpha x(t) = \frac{x(a)}{\Gamma(1-\alpha)}(t-a)^{-\alpha} + \frac{\dot{x}(a)}{\Gamma(2-\alpha)}(t-a)^{1-\alpha}$$
$$+ \frac{(t-a)^{1-\alpha}}{\Gamma(2-\alpha)}\int_a^t \left(\sum_{p=0}^{N}\frac{\Gamma(p-1+\alpha)}{\Gamma(\alpha-1)p!}\left(\frac{\tau-a}{t-a}\right)^p + R_N(\tau)\right)\ddot{x}(\tau)d\tau$$
$$= \frac{x(a)}{\Gamma(1-\alpha)}(t-a)^{-\alpha} + \frac{\dot{x}(a)}{\Gamma(2-\alpha)}(t-a)^{1-\alpha}$$
$$+ \frac{(t-a)^{1-\alpha}}{\Gamma(2-\alpha)}\int_a^t \left(\sum_{p=0}^{N}\frac{\Gamma(p-1+\alpha)}{\Gamma(\alpha-1)p!}\left(\frac{\tau-a}{t-a}\right)^p\right)\ddot{x}(\tau)d\tau$$
$$+ \frac{(t-a)^{1-\alpha}}{\Gamma(2-\alpha)}\int_a^t R_N(\tau)\ddot{x}(\tau)d\tau.$$

At this point, we apply the techniques of (Atanacković and Stanković, 2008) to the first three terms with finite sums. Then, we receive (6.10) with an extra term of truncation error:

$$E_{tr}(t) = \frac{(t-a)^{1-\alpha}}{\Gamma(2-\alpha)}\int_a^t R_N(\tau)\ddot{x}(\tau)d\tau.$$

Since $0 \le \frac{\tau-a}{t-a} \le 1$ for $\tau \in [a,t]$, one has

$$|R_N(\tau)| \le \sum_{p=N+1}^{\infty}\left|\frac{\Gamma(p-1+\alpha)}{\Gamma(\alpha-1)p!}\right| = \sum_{p=N+1}^{\infty}\left|\binom{1-\alpha}{p}\right|$$
$$\le \sum_{p=N+1}^{\infty}\frac{e^{(1-\alpha)^2+1-\alpha}}{p^{2-\alpha}}$$
$$\le \int_{p=N}^{\infty}\frac{e^{(1-\alpha)^2+1-\alpha}}{p^{2-\alpha}}dp = \frac{e^{(1-\alpha)^2+1-\alpha}}{(1-\alpha)N^{1-\alpha}}.$$

Finally, assuming $L_n = \max_{\tau \in [a,t]}\left|x^{(n)}(\tau)\right|$, we conclude that

$$|E_{tr}(t)| \le L_2\frac{e^{(1-\alpha)^2+1-\alpha}}{\Gamma(2-\alpha)(1-\alpha)N^{1-\alpha}}(t-a)^{2-\alpha}.$$

In the general case, the error is given by the following result.

Theorem 6.6. *If we approximate the left Riemann–Liouville fractional derivative by the finite sum (6.17), then the error E_{tr} is bounded by*

$$|E_{tr}(t)| \leq L_n \frac{e^{(n-1-\alpha)^2+n-1-\alpha}}{\Gamma(n-\alpha)(n-1-\alpha)N^{n-1-\alpha}}(t-a)^{n-\alpha}. \qquad (6.30)$$

From (6.30) we see that if the test function grows very fast or the point t is far from a, then the value of N should also increase in order to have a good approximation. Clearly, if we increase the value of n, then we need also to increase the value of N to control the error.

Remark 6.12. Following similar techniques, one can extract an error bound for the approximations of Hadamard derivatives. When we consider finite sums as in (6.24), the error is bounded by

$$|E_{tr}(t)| \leq L(t) \frac{e^{(1-\alpha)^2+1-\alpha}}{\Gamma(2-\alpha)(1-\alpha)N^{1-\alpha}} \left(\ln \frac{t}{a} \right)^{1-\alpha} (t-a),$$

where

$$L(t) = \max_{\tau \in [a,t]} |\dot{x}(\tau) + \tau \ddot{x}(\tau)|.$$

For the general case, the expansion up to the derivative of order n, the error is bounded by

$$|E_{tr}(t)| \leq L_n(t) \frac{e^{(n-\alpha)^2+n-\alpha}}{\Gamma(n+1-\alpha)(n-\alpha)N^{n-\alpha}} \left(\ln \frac{t}{a} \right)^{n-\alpha} (t-a),$$

where

$$L_n(t) = \max_{\tau \in [a,t]} |x_{n,1}(\tau)|.$$

Chapter 7

Approximating fractional integrals

We obtain a new decomposition of the Riemann–Liouville operators of fractional integration as a series involving derivatives (of integer-order). The new formulas are valid for functions of class C^n, $n \in \mathbb{N}$, and allow us to develop suitable numerical approximations with known estimations for the error. The usefulness of the obtained results, in solving fractional integral equations, is illustrated in (Pooseh, Almeida and Torres, 2012a).

7.1 Riemann–Liouville fractional integral

7.1.1 *Approximation by a sum of integer-order derivatives*

For analytical functions, we can rewrite the left Riemann–Liouville fractional integral as a series involving integer-order derivatives only. If x is analytic in $[a, b]$, then

$$_aI_t^\alpha x(t) = \frac{1}{\Gamma(\alpha)} \sum_{k=0}^{\infty} \frac{(-1)^k (t-a)^{k+\alpha}}{(k+\alpha)k!} x^{(k)}(t) \qquad (7.1)$$

for all $t \in [a, b]$ (cf. Eq. (3.44) in (Miller and Ross, 1993)). From the numerical point of view, one considers finite sums and the following approximation:

$$_aI_t^\alpha x(t) \approx \frac{1}{\Gamma(\alpha)} \sum_{k=0}^{N} \frac{(-1)^k (t-a)^{k+\alpha}}{(k+\alpha)k!} x^{(k)}(t). \qquad (7.2)$$

One problem with formula (7.1) is that in order to have a "good" approximation we need to take a large value for n. In applications, this approach may not be suitable. Here we present a new decomposition formula for functions of class C^n. The advantage is that even for $n = 1$, we can achieve an appropriate accuracy.

7.1.2　*Approximation using moments of a function*

Before we give the result in its full extension, we explain the method for $n = 3$. To that purpose, let $x \in C^3[a, b]$. Using integration by parts three times, we deduce that

$$
{}_aI_t^\alpha x(t) = \frac{x(a)}{\Gamma(\alpha+1)}(t-a)^\alpha + \frac{\dot{x}(a)}{\Gamma(\alpha+2)}(t-a)^{\alpha+1} + \frac{\ddot{x}(a)}{\Gamma(\alpha+3)}(t-a)^{\alpha+2}
$$
$$
+ \frac{1}{\Gamma(\alpha+3)} \int_a^t (t-\tau)^{\alpha+2} x^{(3)}(\tau)d\tau.
$$

By the binomial formula, we can rewrite the fractional integral as

$$
{}_aI_t^\alpha x(t) = \frac{x(a)}{\Gamma(\alpha+1)}(t-a)^\alpha + \frac{\dot{x}(a)}{\Gamma(\alpha+2)}(t-a)^{\alpha+1} + \frac{\ddot{x}(a)}{\Gamma(\alpha+3)}(t-a)^{\alpha+2}
$$
$$
+ \frac{(t-a)^{\alpha+2}}{\Gamma(\alpha+3)} \sum_{p=0}^\infty \frac{\Gamma(p-\alpha-2)}{\Gamma(-\alpha-2)p!(t-a)^p} \int_a^t (\tau-a)^p x^{(3)}(\tau)d\tau.
$$

The rest of the procedure follows the same pattern: decompose the sum into a first term plus the others, and integrate by parts. Then assuming

$$
A_0(\alpha) = \frac{1}{\Gamma(\alpha+1)}\left[1 + \sum_{p=3}^\infty \frac{\Gamma(p-\alpha-2)}{\Gamma(-\alpha)(p-2)!}\right],
$$
$$
A_1(\alpha) = \frac{1}{\Gamma(\alpha+2)}\left[1 + \sum_{p=2}^\infty \frac{\Gamma(p-\alpha-2)}{\Gamma(-\alpha-1)(p-1)!}\right],
$$
$$
A_2(\alpha) = \frac{1}{\Gamma(\alpha+3)}\left[1 + \sum_{p=1}^\infty \frac{\Gamma(p-\alpha-2)}{\Gamma(-\alpha-2)p!}\right],
$$

we obtain

$$
{}_aI_t^\alpha x(t) = \frac{x(a)}{\Gamma(\alpha+1)}(t-a)^\alpha + \frac{\dot{x}(a)}{\Gamma(\alpha+2)}(t-a)^{\alpha+1} + A_2(\alpha)(t-a)^{\alpha+2}\ddot{x}(t)
$$
$$
+ \frac{(t-a)^{\alpha+2}}{\Gamma(\alpha+2)} \sum_{p=1}^\infty \frac{\Gamma(p-\alpha-2)}{\Gamma(-\alpha-1)(p-1)!(t-a)^p} \int_a^t (\tau-a)^{p-1}\ddot{x}(\tau)d\tau
$$
$$
= \frac{x(a)}{\Gamma(\alpha+1)}(t-a)^\alpha + A_1(\alpha)(t-a)^{\alpha+1}\dot{x}(t) + A_2(\alpha)(t-a)^{\alpha+2}\ddot{x}(t)
$$
$$
+ \frac{(t-a)^{\alpha+2}}{\Gamma(\alpha+1)} \sum_{p=2}^\infty \frac{\Gamma(p-\alpha-2)}{\Gamma(-\alpha)(p-2)!(t-a)^p} \int_a^t (\tau-a)^{p-2}\dot{x}(\tau)d\tau
$$
$$
= A_0(\alpha)(t-a)^\alpha x(t) + A_1(\alpha)(t-a)^{\alpha+1}\dot{x}(t) + A_2(\alpha)(t-a)^{\alpha+2}\ddot{x}(t)
$$
$$
+ \frac{(t-a)^{\alpha+2}}{\Gamma(\alpha)} \sum_{p=3}^\infty \frac{\Gamma(p-\alpha-2)}{\Gamma(-\alpha+1)(p-3)!(t-a)^p} \int_a^t (\tau-a)^{p-3}x(\tau)d\tau.
$$

Therefore, we can expand $_aI_t^\alpha x(t)$ as

$$_aI_t^\alpha x(t) = A_0(\alpha)(t-a)^\alpha x(t) + A_1(\alpha)(t-a)^{\alpha+1}\dot{x}(t)$$

$$+A_2(\alpha)(t-a)^{\alpha+2}\ddot{x}(t) + \sum_{p=3}^\infty B(\alpha,p)(t-a)^{\alpha+2-p}V_p(t), \qquad (7.3)$$

where

$$B(\alpha,p) = \frac{\Gamma(p-\alpha-2)}{\Gamma(\alpha)\Gamma(1-\alpha)(p-2)!}, \qquad (7.4)$$

and

$$V_p(t) = \int_a^t (p-2)(\tau-a)^{p-3}x(\tau)d\tau. \qquad (7.5)$$

Remark 7.1. Function V_p given by (7.5) may be defined as the solution of the differential equation

$$\begin{cases} \dot{V}_p(t) = (p-2)(t-a)^{p-3}x(t) \\ V_p(a) = 0, \end{cases}$$

for $p = 3, 4, \ldots$

Remark 7.2. When α is not an integer, we may use Euler's reflection formula (cf. (Beals and Wong, 2010))

$$\Gamma(\alpha)\Gamma(1-\alpha) = \frac{\pi}{\sin(\pi\alpha)},$$

to simplify expression $B(\alpha,p)$ in (7.4).

Following the same reasoning, we are able to deduce a general formula of decomposition for fractional integrals, depending on the order of smoothness of the test function.

Theorem 7.1. *Let $n \in \mathbb{N}$ and $x \in C^n[a,b]$. Then*

$$_aI_t^\alpha x(t) = \sum_{i=0}^{n-1} A_i(\alpha)(t-a)^{\alpha+i}x^{(i)}(t) + \sum_{p=n}^\infty B(\alpha,p)(t-a)^{\alpha+n-1-p}V_p(t),$$

$$(7.6)$$

where

$$A_i(\alpha) = \frac{1}{\Gamma(\alpha+i+1)}\left[1 + \sum_{p=n-i}^\infty \frac{\Gamma(p-\alpha-n+1)}{\Gamma(-\alpha-i)(p-n+1+i)!}\right],$$

$i = 0, \ldots, n-1,$

$$B(\alpha,p) = \frac{\Gamma(p-\alpha-n+1)}{\Gamma(\alpha)\Gamma(1-\alpha)(p-n+1)!}, \qquad (7.7)$$

and

$$V_p(t) = \int_a^t (p - n + 1)(\tau - a)^{p-n} x(\tau) d\tau, \tag{7.8}$$

$p = n, n+1, \ldots$

A remark about the convergence of the series in $A_i(\alpha)$, $i \in \{0, \ldots, n-1\}$, is in order. Since

$$\sum_{p=n-i}^{\infty} \frac{\Gamma(p - \alpha - n + 1)}{\Gamma(-\alpha - i)(p - n + 1 + i)!} = \sum_{p=0}^{\infty} \frac{\Gamma(p - \alpha - i)}{\Gamma(-\alpha - i)p!} - 1 \tag{7.9}$$
$$= {}_1F_0(-\alpha - i, 1),$$

where ${}_1F_0$ denotes the hypergeometric function, and because $\alpha + i > 0$, we conclude that (7.9) converges absolutely (cf. Theorem 2.1.2 in (Andrews, Askey and Roy, 1999)). In fact, we may use Eq. (2.1.6) in (Andrews, Askey and Roy, 1999) to conclude that

$$\sum_{p=n-i}^{\infty} \frac{\Gamma(p - \alpha - n + 1)}{\Gamma(-\alpha - i)(p - n + 1 + i)!} = -1.$$

Therefore, the first n terms of our decomposition (7.6) vanish. However, because of numerical reasons, we do not follow this procedure here. Indeed, only finite sums of these coefficients are to be taken, and we obtain a better accuracy for the approximation taking them into account (see Figs. 7.5(a) and 7.5(b)). More precisely, we consider finite sums up to order N, with $N \geq n$. Thus, our approximation will depend on two parameters: the order of the derivative $n \in \mathbb{N}$, and the number of terms taken in the sum, which is given by N. The left fractional integral is then approximated by

$${}_aI_t^\alpha x(t) \approx \sum_{i=0}^{n-1} A_i(\alpha, N)(t - a)^{\alpha+i} x^{(i)}(t) + \sum_{p=n}^{N} B(\alpha, p)(t - a)^{\alpha+n-1-p} V_p(t), \tag{7.10}$$

where

$$A_i(\alpha, N) = \frac{1}{\Gamma(\alpha + i + 1)} \left[1 + \sum_{p=n-i}^{N} \frac{\Gamma(p - \alpha - n + 1)}{\Gamma(-\alpha - i)(p - n + 1 + i)!} \right], \tag{7.11}$$

and $B(\alpha, p)$ and $V_p(t)$ are given by (7.7) and (7.8), respectively.

To measure the truncation errors made by neglecting the remaining terms, observe that

$$\frac{1}{\Gamma(\alpha+i+1)} \sum_{p=N+1}^{\infty} \frac{\Gamma(p-\alpha-n+1)}{\Gamma(-\alpha-i)(p-n+1+i)!}$$

$$= \frac{1}{\Gamma(\alpha+i+1)} \sum_{p=N-n+2+i}^{\infty} \frac{\Gamma(p-\alpha-i)}{\Gamma(-\alpha-i)p!}$$

$$= \frac{1}{\Gamma(\alpha+i+1)} \left[{}_2F_1(-\alpha-i,-,-,1) - \sum_{p=0}^{N-n+1+i} \frac{\Gamma(p-\alpha-i)}{\Gamma(-\alpha-i)p!} \right]$$

$$= \frac{-1}{\Gamma(\alpha+i+1)} \sum_{p=0}^{N-n+i+1} \frac{\Gamma(p-\alpha-i)}{\Gamma(-\alpha-i)p!}.$$

$$(7.12)$$

Similarly,

$$\frac{1}{\Gamma(\alpha)\Gamma(1-\alpha)} \sum_{p=N+1}^{\infty} \frac{\Gamma(p-\alpha-n+1)}{(p-n+1)!}$$

$$= \frac{-1}{\Gamma(\alpha)\Gamma(1-\alpha)} \sum_{p=0}^{N-n+1} \frac{\Gamma(p-\alpha)}{p!}.$$

$$(7.13)$$

In Tables 7.1 and 7.2 we exemplify some values for (7.12) and (7.13), respectively, with $\alpha = 0.5$ and for different values of N, n and i. Observe that the errors only depend on the values of $N - n$ and i for (7.12), and on the value of $N - n$ for (7.13).

Table 7.1   Values of error (7.12) for $\alpha = 0.5$.

i \ $N-n$	0	1	2	3	4
0	-0.5642	-0.4231	-0.3526	-0.3085	-0.2777
1	0.09403	0.04702	0.02938	0.02057	0.01543
2	-0.01881	-0.007052	-0.003526	-0.002057	-0.001322
3	0.003358	0.001007	0.0004198	0.0002099	0.0001181
4	-0.0005224	-0.0001306	-0.00004664	-0.00002041	-0.00001020
5	7.12×10^{-5}	1.52×10^{-5}	4.77×10^{-6}	1.85×10^{-6}	8.34×10^{-7}

Everything done so far is easily adapted to the right fractional integral. In fact, one has:

Table 7.2   Values of error (7.13) for $\alpha = 0.5$.

$N - n$	0	1	2	3	4
	0.5642	0.4231	0.3526	0.3085	0.2777

Theorem 7.2. *Let* $n \in \mathbb{N}$ *and* $x \in C^n[a, b]$. *Then*

$$
{}_t I_b^\alpha x(t) = \sum_{i=0}^{n-1} A_i(\alpha)(b-t)^{\alpha+i} x^{(i)}(t) + \sum_{p=n}^\infty B(\alpha, p)(b-t)^{\alpha+n-1-p} W_p(t),
$$

where

$$
A_i(\alpha) = \frac{(-1)^i}{\Gamma(\alpha + i + 1)} \left[1 + \sum_{p=n-i}^\infty \frac{\Gamma(p - \alpha - n + 1)}{\Gamma(-\alpha - i)(p - n + 1 + i)!} \right],
$$

$$
B(\alpha, p) = \frac{(-1)^n \Gamma(p - \alpha - n + 1)}{\Gamma(\alpha)\Gamma(1 - \alpha)(p - n + 1)!},
$$

$$
W_p(t) = \int_t^b (p - n + 1)(b - \tau)^{p-n} x(\tau) d\tau.
$$

7.1.3   *Numerical evaluation of fractional integrals*

In this section we exemplify the proposed approximation procedure with some examples. In each step, we evaluate the accuracy of our method, i.e., the error when substituting ${}_a I_t^\alpha x$ by an approximation ${}_a \tilde{I}_t^\alpha x$. For that purpose, we take the distance given by

$$
E = \sqrt{\int_a^b \left({}_a I_t^\alpha x(t) - {}_a \tilde{I}_t^\alpha x(t) \right)^2 dt}.
$$

Firstly, consider $x_1(t) = t^3$ and $x_2(t) = t^{10}$ with $t \in [0, 1]$. Then

$$
{}_0 I_t^{0.5} x_1(t) = \frac{\Gamma(4)}{\Gamma(4.5)} t^{3.5} \quad \text{and} \quad {}_0 I_t^{0.5} x_2(t) = \frac{\Gamma(11)}{\Gamma(11.5)} t^{10.5}
$$

(cf. Property 2.1 in (Kilbas, Srivastava and Trujillo, 2006)). Let us consider Theorem 7.1 for $n = 3$, i.e., expansion (7.3) for different values of step N. For function x_1, small values of N are enough ($N = 3, 4, 5$). For x_2 we take $N = 4, 6, 8$. In Figs. 7.1(a) and 7.1(b) we represent the graphs of the fractional integrals of x_1 and x_2 of order $\alpha = 0.5$ together with different approximations. As expected, when N increases we obtain a better approximation for each fractional integral.

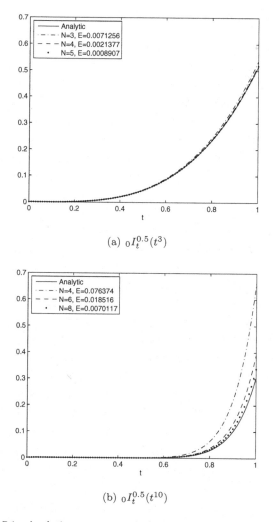

(a) $_0I_t^{0.5}(t^3)$

(b) $_0I_t^{0.5}(t^{10})$

Fig. 7.1   Analytic *versus* numerical approximation for a fixed n.

Secondly, we apply our procedure to the transcendental functions $x_3(t) = e^t$ and $x_4(t) = \sin(t)$. Simple calculations give

$$_0I_t^{0.5}x_3(t) = \sqrt{t}\sum_{k=0}^{\infty}\frac{t^k}{\Gamma(k+1.5)} \quad \text{and} \quad _0I_t^{0.5}x_4(t) = \sqrt{t}\sum_{k=0}^{\infty}\frac{(-1)^k t^{2k+1}}{\Gamma(2k+2.5)}.$$

Figures 7.2(a) and 7.2(b) show the numerical results for each approximation, with $n = 3$. We see that for a small value of N one already obtains a good approximation for each function.

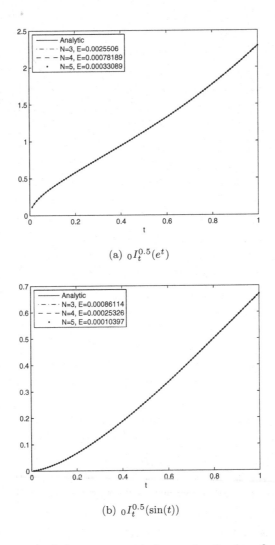

(a) $_0I_t^{0.5}(e^t)$

(b) $_0I_t^{0.5}(\sin(t))$

Fig. 7.2 Analytic *versus* numerical approximation for a fixed n.

For analytical functions, we may apply the well-known formula (7.2). In Fig. 7.3 we show the results of approximating with (7.2), $N = 1, 2, 3$, for functions $x_3(t)$ and $x_4(t)$. We remark that, when we consider expansions up to the second derivative, i.e., the cases $n = 3$ as in (7.3) and expansion

(7.2) with $N = 2$, we obtain a better accuracy using our approximation (7.3) even for a small value of N.

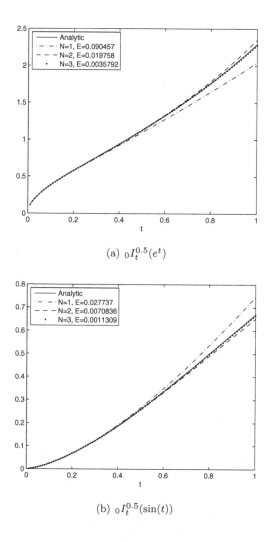

(a) $_0I_t^{0.5}(e^t)$

(b) $_0I_t^{0.5}(\sin(t))$

Fig. 7.3   Numerical approximation using (7.2) of previous literature.

Another way to approximate fractional integrals is to fix N and consider several sizes for the decomposition, i.e., letting n to vary. Let us consider the two test functions $x_1(t) = t^3$ and $x_2(t) = t^{10}$, with $t \in [0, 1]$ as before.

In both cases we consider the first three approximations of the fractional integral, i.e., for $n = 1, 2, 3$. For the first function we fix $N = 3$, for the second one we choose $N = 8$. Figures 7.4(a) and 7.4(b) show the numerical results. As expected, for a greater value of n the error decreases.

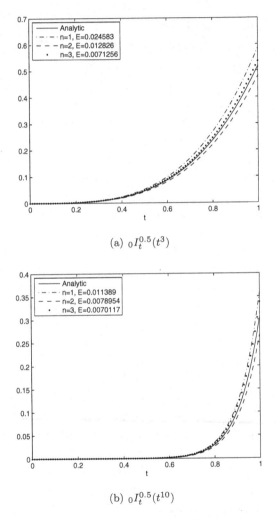

(a) $_0I_t^{0.5}(t^3)$

(b) $_0I_t^{0.5}(t^{10})$

Fig. 7.4   Analytic *versus* numerical approximation for a fixed N.

We mentioned before that although the terms A_i are all equal to zero, for $i \in \{0, \ldots, n-1\}$, we consider them in the decomposition formula. Indeed, after we truncate the sum, the error is lower. This is illustrated in Figs. 7.5(a) and 7.5(b), where we study the approximations for $_0I_t^{0.5}x_1(t)$ and $_0I_t^{0.5}x_2(t)$ with $A_i \neq 0$ and $A_i = 0$.

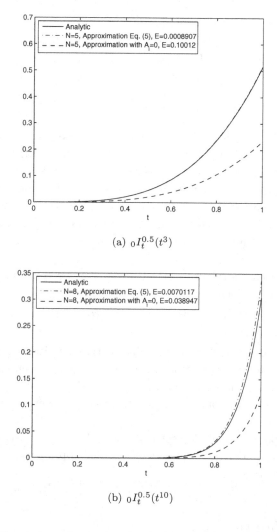

(a) $_0I_t^{0.5}(t^3)$

(b) $_0I_t^{0.5}(t^{10})$

Fig. 7.5 Comparison of approximation (7.3) and approximation with $A_i = 0$.

7.1.4 Applications to fractional integral equations

In this section we show how the proposed approximations can be applied to solve a fractional integral equation (Example 7.1), which depends on the left Riemann–Liouville fractional integral. The main idea is to rewrite the initial problem by replacing the fractional integrals by an expansion of type (7.1) or (7.6), thus getting a problem involving integer-order derivatives, which can be solved by standard techniques.

Example 7.1. Consider the following fractional system:

$$\begin{cases} {}_0I_t^{0.5}x(t) = \frac{\Gamma(4.5)}{24}t^4 \\ x(0) = 0. \end{cases} \tag{7.14}$$

Since ${}_0I_t^{0.5}t^{3.5} = \frac{\Gamma(4.5)}{24}t^\alpha$, the function $t \mapsto t^{3.5}$ is a solution to problem (7.14).

To provide a numerical method to solve such type of systems, we replace the fractional integral by approximations (7.2) and (7.10), for a suitable order. We remark that the order of approximation, N in (7.2) and n in (7.10), are restricted by the number of given initial or boundary conditions. Since (7.14) has one initial condition, in order to solve it numerically, we will consider the expansion for the fractional integral up to the first derivative, i.e., $N = 1$ in (7.2) and $n = 2$ in (7.10). The order N in (7.10) can be freely chosen.

Applying approximation (7.2), with $\alpha = 0.5$, we transform (7.14) into the initial value problem

$$\begin{cases} 1.1285t^{0.5}x(t) - 0.3761t^{1.5}\dot{x}(t) = \frac{\Gamma(4.5)}{24}t^4, \\ x(0) = 0, \end{cases}$$

which is a first order ODE (ordinary differential equation). The solution is shown in Fig. 7.6(a). It reveals that the approximation remains close to the exact solution for a short time and diverges drastically afterwards. Since we have no extra information, we cannot increase the order of approximation to proceed.

To use expansion (7.6), we rewrite the problem as a standard one, depending only on a derivative of first order. The approximated system that we must solve is

$$\begin{cases} A_0(0.5, N)t^{0.5}x(t) + A_1(0.5, N)t^{1.5}\dot{x}(t) + \sum_{p=2}^N B(0.5, p)t^{1.5-p}V_p(t) \\ \quad = \frac{\Gamma(4.5)}{24}t^4, \\ \dot{V}_p(t) = (p-1)t^{p-2}x(t), \quad p = 2, 3, \ldots, N, \\ x(0) = 0, \\ V_p(0) = 0, \quad p = 2, 3, \ldots, N, \end{cases}$$

where A_0 and A_1 are given as in (7.11) and B is given by Theorem 7.1. Here, by increasing N, we get better approximations to the fractional integral and we expect more accurate solutions to the original problem (7.14). For $N = 2$ and $N = 3$ we transform the resulting system of ordinary differential equations to a second and a third-order differential equation, respectively. Finally, we solve them using the Maple® built in function dsolve. For example, for $N = 2$ the second-order equation takes the form

$$\begin{cases} \ddot{V}_2(t) = \frac{6}{t}\dot{V}_2(t) + \frac{6}{t^2}V_2(t) - 5.1542t^{2.5} \\ V_2(0) = 0 \\ \dot{V}_2(0) = x(0) = 0, \end{cases}$$

and the solution is $x(t) = \dot{V}_2(t) = 1.34t^{3.5}$. In Fig. 7.6(b) we compare the exact solution with numerical approximations for two values of N.

7.2 Hadamard fractional integrals

7.2.1 *Approximation by a sum of integer-order derivatives*

For an arbitrary $\alpha > 0$ we refer the reader to (Kilbas, 2001, Theorem 3.2). If a function x admits derivatives of any order, then expansion formulas for the Hadamard fractional integrals and derivatives of x, in terms of its integer-order derivatives, are given in (Butzer, Kilbas and Trujillo, 2003, Theorem 17):

$$_0\mathcal{I}_t^\alpha x(t) = \sum_{k=0}^\infty S(-\alpha, k)t^k x^{(k)}(t)$$

and

$$_0\mathcal{D}_t^\alpha x(t) = \sum_{k=0}^\infty S(\alpha, k)t^k x^{(k)}(t),$$

where

$$S(\alpha, k) = \frac{1}{k!}\sum_{j=1}^k (-1)^{k-j}\binom{k}{j}j^\alpha$$

is the Stirling function.

7.2.2 *Approximation using moments of a function*

In this section we consider the class of differentiable functions up to order $n + 1$, $x \in C^{n+1}[a, b]$, and deduce expansion formulas for the Hadamard

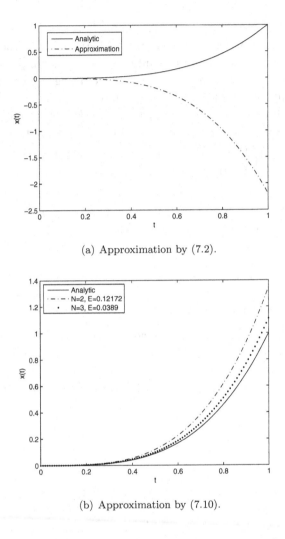

(a) Approximation by (7.2).

(b) Approximation by (7.10).

Fig. 7.6   Analytic *versus* numerical solution to problem (7.14).

fractional integrals in terms of $x^{(i)}$, for $i \in \{0, \ldots, n\}$. Before presenting the result in its full extension, we briefly explain the techniques involved for the particular case $n = 2$. To that purpose, let $x \in C^3[a, b]$. Integrating by parts three times, we obtain

$$
\begin{aligned}
_a\mathcal{I}_t^\alpha x(t) &= -\frac{1}{\Gamma(\alpha)} \int_a^t -\frac{1}{\tau} \left(\ln \frac{t}{\tau} \right)^{\alpha-1} x(\tau) d\tau \\
&= \frac{1}{\Gamma(\alpha+1)} \left(\ln \frac{t}{a} \right)^\alpha x(a) - \frac{1}{\Gamma(\alpha+1)} \int_a^t -\frac{1}{\tau} \left(\ln \frac{t}{\tau} \right)^\alpha \tau \dot{x}(\tau) d\tau \\
&= \frac{1}{\Gamma(\alpha+1)} \left(\ln \frac{t}{a} \right)^\alpha x(a) + \frac{1}{\Gamma(\alpha+2)} \left(\ln \frac{t}{a} \right)^{\alpha+1} a\dot{x}(a) \\
&\quad - \frac{1}{\Gamma(\alpha+2)} \int_a^t -\frac{1}{\tau} \left(\ln \frac{t}{\tau} \right)^{\alpha+1} (\tau \dot{x}(\tau) + \tau^2 \ddot{x}(\tau)) d\tau \\
&= \frac{1}{\Gamma(\alpha+1)} \left(\ln \frac{t}{a} \right)^\alpha x(a) + \frac{1}{\Gamma(\alpha+2)} \left(\ln \frac{t}{a} \right)^{\alpha+1} a\dot{x}(a) \\
&\quad + \frac{1}{\Gamma(\alpha+3)} \left(\ln \frac{t}{a} \right)^{\alpha+2} (a\dot{x}(a) + a^2 \ddot{x}(a)) \\
&\quad + \frac{1}{\Gamma(\alpha+3)} \int_a^t \left(\ln \frac{t}{\tau} \right)^{\alpha+2} (\dot{x}(\tau) + 3\tau \ddot{x}(\tau) + \tau^2 \dddot{x}(\tau)) d\tau.
\end{aligned}
$$

On the other hand, using the binomial theorem, we have

$$
\begin{aligned}
\left(\ln \frac{t}{\tau} \right)^{\alpha+2} &= \left(\ln \frac{t}{a} \right)^{\alpha+2} \left(1 - \frac{\ln \frac{\tau}{a}}{\ln \frac{t}{a}} \right)^{\alpha+2} \\
&= \left(\ln \frac{t}{a} \right)^{\alpha+2} \sum_{p=0}^\infty \frac{\Gamma(p-\alpha-2)}{\Gamma(-\alpha-2)p!} \cdot \frac{(\ln \frac{\tau}{a})^p}{(\ln \frac{t}{a})^p}.
\end{aligned}
$$

This series converges since $\tau \in [a, t]$ and $\alpha + 2 > 0$. Combining these formulas, we get

$$
\begin{aligned}
_a\mathcal{I}_t^\alpha x(t) &= \frac{x(a)}{\Gamma(\alpha+1)} \left(\ln \frac{t}{a} \right)^\alpha \\
&+ \frac{a\dot{x}(a)}{\Gamma(\alpha+2)} \left(\ln \frac{t}{a} \right)^{\alpha+1} + \frac{a\dot{x}(a) + a^2 \ddot{x}(a)}{\Gamma(\alpha+3)} \left(\ln \frac{t}{a} \right)^{\alpha+2} \\
&+ \frac{(\ln \frac{t}{a})^{\alpha+2}}{\Gamma(\alpha+3)} \sum_{p=0}^\infty \Gamma_0(\alpha, p, t) \int_a^t \left(\ln \frac{\tau}{a} \right)^p (\dot{x}(\tau) + 3\tau \ddot{x}(\tau) + \tau^2 \dddot{x}(\tau)) \, d\tau,
\end{aligned}
$$

where

$$
\Gamma_i(\alpha, p, t) = \frac{\Gamma(p-\alpha-2)}{\Gamma(-\alpha-2+i)(p-i)! \left(\ln \frac{t}{a} \right)^p}.
$$

Now, split the series into the two cases $p = 0$ and $p = 1 \ldots \infty$, and integrate by parts the second one. We obtain

$$
_aI_t^\alpha x(t) = \frac{1}{\Gamma(\alpha+1)} \left(\ln \frac{t}{a}\right)^\alpha x(a) + \frac{1}{\Gamma(\alpha+2)} \left(\ln \frac{t}{a}\right)^{\alpha+1} a\dot{x}(a)
$$
$$
+ \frac{1}{\Gamma(\alpha+3)} \left(\ln \frac{t}{a}\right)^{\alpha+2} (t\dot{x}(t) + t^2\ddot{x}(t)) \left[1 + \sum_{p=1}^{\infty} \frac{\Gamma(p-\alpha-2)}{\Gamma(-\alpha-2)p!}\right]
$$
$$
+ \frac{\left(\ln \frac{t}{a}\right)^{\alpha+2}}{\Gamma(\alpha+2)} \sum_{p=1}^{\infty} \Gamma_1(\alpha, p, t) \int_a^t \left(\ln \frac{\tau}{a}\right)^{p-1} (\dot{x}(\tau) + \tau\ddot{x}(\tau))d\tau.
$$

Repeating this procedure two more times, we obtain the following:

$$
_aI_t^\alpha x(t) = \frac{1}{\Gamma(\alpha+1)} \left(\ln \frac{t}{a}\right)^\alpha x(t) \left[1 + \sum_{p=3}^{\infty} \frac{\Gamma(p-\alpha-2)}{\Gamma(-\alpha)(p-2)!}\right]
$$
$$
+ \frac{1}{\Gamma(\alpha+2)} \left(\ln \frac{t}{a}\right)^{\alpha+1} t\dot{x}(t) \left[1 + \sum_{p=2}^{\infty} \frac{\Gamma(p-\alpha-2)}{\Gamma(-\alpha-1)(p-1)!}\right]
$$
$$
+ \frac{1}{\Gamma(\alpha+3)} \left(\ln \frac{t}{a}\right)^{\alpha+2} (t\dot{x}(t) + t^2\ddot{x}(t)) \left[1 + \sum_{p=1}^{\infty} \frac{\Gamma(p-\alpha-2)}{\Gamma(-\alpha-2)p!}\right]
$$
$$
+ \frac{\left(\ln \frac{t}{a}\right)^{\alpha+2}}{\Gamma(\alpha)} \sum_{p=3}^{\infty} \frac{\Gamma(p-\alpha-2)}{\Gamma(-\alpha+1)(p-3)!\left(\ln \frac{t}{a}\right)^p} \int_a^t \left(\ln \frac{\tau}{a}\right)^{p-3} \frac{x(\tau)}{\tau} d\tau,
$$

or, in a more concise way,

$$
_aI_t^\alpha x(t) = A_0(\alpha) \left(\ln \frac{t}{a}\right)^\alpha x(t) + A_1(\alpha) \left(\ln \frac{t}{a}\right)^{\alpha+1} t\dot{x}(t)
$$
$$
+ A_2(\alpha) \left(\ln \frac{t}{a}\right)^{\alpha+2} (t\dot{x}(t) + t^2\ddot{x}(t)) + \sum_{p=3}^{\infty} B(\alpha, p) \left(\ln \frac{t}{a}\right)^{\alpha+2-p} V_p(t),
$$

with

$$
A_0(\alpha) = \frac{1}{\Gamma(\alpha+1)} \left[1 + \sum_{p=3}^{\infty} \frac{\Gamma(p-\alpha-2)}{\Gamma(-\alpha)(p-2)!}\right],
$$
$$
A_1(\alpha) = \frac{1}{\Gamma(\alpha+2)} \left[1 + \sum_{p=2}^{\infty} \frac{\Gamma(p-\alpha-2)}{\Gamma(-\alpha-1)(p-1)!}\right],
$$
$$
A_2(\alpha) = \frac{1}{\Gamma(\alpha+3)} \left[1 + \sum_{p=1}^{\infty} \frac{\Gamma(p-\alpha-2)}{\Gamma(-\alpha-2)p!}\right],
$$

$$
B(\alpha, p) = \frac{\Gamma(p-\alpha-2)}{\Gamma(\alpha)\Gamma(1-\alpha)(p-2)!}, \tag{7.15}
$$

and

$$V_p(t) = \int_a^t (p-2)\left(\ln\frac{\tau}{a}\right)^{p-3}\frac{x(\tau)}{\tau}\,d\tau, \tag{7.16}$$

where we assume the series and the integral V_p to be convergent.

Remark 7.3. When useful, namely on fractional differential and integral equations, we can define V_p as in (7.16) by the solution of the system

$$\begin{cases} \dot{V}_p(t) = (p-2)\left(\ln\frac{t}{a}\right)^{p-3}\frac{x(t)}{t} \\ V_p(a) = 0, \end{cases}$$

for all $p = 3, 4, \ldots$

We now discuss the convergence of the series involved in the definitions of $A_i(\alpha)$, for $i \in \{0, 1, 2\}$. Simply observe that

$$\sum_{p=3-i}^{\infty} \frac{\Gamma(p-\alpha-2)}{\Gamma(-\alpha-i)(p-2+i)!} = {}_1F_0(-\alpha-i, 1) - 1,$$

and ${}_1F_0(a, x)$ converges absolutely when $|x| = 1$ if $a < 0$ (Andrews, Askey and Roy, 1999, Theorem 2.1.2).

For numerical purposes, only finite sums are considered, and thus the Hadamard left fractional integral is approximated by the decomposition

$$_a\mathcal{I}_t^\alpha x(t) \approx A_0(\alpha, N)\left(\ln\frac{t}{a}\right)^\alpha x(t) + A_1(\alpha, N)\left(\ln\frac{t}{a}\right)^{\alpha+1} t\dot{x}(t)$$

$$+ A_2(\alpha, N)\left(\ln\frac{t}{a}\right)^{\alpha+2}(t\dot{x}(t) + t^2\ddot{x}(t)) + \sum_{p=3}^{N} B(\alpha, p)\left(\ln\frac{t}{a}\right)^{\alpha+2-p} V_p(t),$$

$$\tag{7.17}$$

with

$$A_0(\alpha, N) = \frac{1}{\Gamma(\alpha+1)}\left[1 + \sum_{p=3}^{N}\frac{\Gamma(p-\alpha-2)}{\Gamma(-\alpha)(p-2)!}\right],$$

$$A_1(\alpha, N) = \frac{1}{\Gamma(\alpha+2)}\left[1 + \sum_{p=2}^{N}\frac{\Gamma(p-\alpha-2)}{\Gamma(-\alpha-1)(p-1)!}\right],$$

$$A_2(\alpha, N) = \frac{1}{\Gamma(\alpha+3)}\left[1 + \sum_{p=1}^{N}\frac{\Gamma(p-\alpha-2)}{\Gamma(-\alpha-2)p!}\right],$$

$B(\alpha, p)$ and $V_p(t)$ as in (7.15) and (7.16), and $N \geq 3$.

Following similar arguments as shown for $n = 2$, we can prove the general case with an expansion up to the derivative of order n. First, we introduce a notation. Given $k \in \mathbb{N} \cup \{0\}$, we define the sequences $x_{k,0}(t)$ and $x_{k,1}(t)$ recursively by the formulas

$$x_{0,0}(t) = x(t) \text{ and } x_{k+1,0}(t) = t\frac{d}{dt}x_{k,0}(t), \text{ for } k \in \mathbb{N} \cup \{0\},$$

and

$$x_{0,1}(t) = \dot{x}(t) \text{ and } x_{k+1,1}(t) = \frac{d}{dt}(tx_{k,1}(t)), \text{ for } k \in \mathbb{N} \cup \{0\}.$$

Theorem 7.3. *Let $n \in \mathbb{N}$, $0 < a < b$ and $x : [a,b] \to \mathbb{R}$ be a function of class C^{n+1}. Then,*

$$_a\mathcal{I}_t^\alpha x(t) = \sum_{i=0}^{n} A_i(\alpha) \left(\ln\frac{t}{a}\right)^{\alpha+i} x_{i,0}(t) + \sum_{p=n+1}^{\infty} B(\alpha, p) \left(\ln\frac{t}{a}\right)^{\alpha+n-p} V_p(t)$$

with

$$A_i(\alpha) = \frac{1}{\Gamma(\alpha + i + 1)}\left[1 + \sum_{p=n-i+1}^{\infty} \frac{\Gamma(p - \alpha - n)}{\Gamma(-\alpha - i)(p - n + i)!}\right],$$

$$B(\alpha, p) = \frac{\Gamma(p - \alpha - n)}{\Gamma(\alpha)\Gamma(1 - \alpha)(p - n)!},$$

$$V_p(t) = \int_a^t (p - n)\left(\ln\frac{\tau}{a}\right)^{p-n-1} \frac{x(\tau)}{\tau} d\tau.$$

Proof. Applying integration by parts repeatedly and the binomial formula, we arrive to

$$_a\mathcal{I}_t^\alpha x(t) = \sum_{i=0}^{n} \frac{1}{\Gamma(\alpha + i + 1)}\left(\ln\frac{t}{a}\right)^{\alpha+i} x_{i,0}(a)$$

$$+ \frac{1}{\Gamma(\alpha + n + 1)}\left(\ln\frac{t}{a}\right)^{\alpha+n} \sum_{p=0}^{\infty} \frac{\Gamma(p - \alpha - n)}{\Gamma(-\alpha - n)p!\left(\ln\frac{t}{a}\right)^p} \int_a^t \left(\ln\frac{\tau}{a}\right)^p x_{n,1}(\tau)d\tau.$$

To achieve the expansion formula, we repeat the same procedure as for the case $n = 2$: we split the sum into two parts (the first term plus the remaining) and integrate by parts the second one. The convergence of the series $A_i(\alpha)$ is ensured by the relation

$$\sum_{p=n-i+1}^{\infty} \frac{\Gamma(p - \alpha - n)}{\Gamma(-\alpha - i)(p - n + i)!} = {}_1F_0(-\alpha - i, 1) - 1.$$

$\square$

An estimation for the error bound is given in Section 7.3.

Similarly to what was done with the left fractional integral, we can also expand the right Hadamard fractional integral.

Theorem 7.4. *Let* $n \in \mathbb{N}$, $0 < a < b$ *and* $x : [a, b] \to \mathbb{R}$ *be a function of class* C^{n+1}. *Then,*

$$_t\mathcal{I}_b^\alpha x(t) = \sum_{i=0}^{n} A_i(\alpha) \left(\ln \frac{b}{t} \right)^{\alpha+i} x_{i,0}(t) + \sum_{p=n+1}^{\infty} B(\alpha, p) \left(\ln \frac{b}{t} \right)^{\alpha+n-p} W_p(t)$$

with

$$A_i(\alpha) = \frac{(-1)^i}{\Gamma(\alpha + i + 1)} \left[1 + \sum_{p=n-i+1}^{\infty} \frac{\Gamma(p - \alpha - n)}{\Gamma(-\alpha - i)(p - n + i)!} \right],$$

$$B(\alpha, p) = \frac{\Gamma(p - \alpha - n)}{\Gamma(\alpha)\Gamma(1 - \alpha)(p - n)!},$$

$$W_p(t) = \int_t^b (p - n) \left(\ln \frac{b}{\tau} \right)^{p-n-1} \frac{x(\tau)}{\tau} d\tau.$$

Remark 7.4. Analogously to what was done for the left fractional integral, one can consider an approximation for the right Hadamard fractional integral by considering finite sums in the expansion obtained in Theorem 7.4.

7.2.3 *Examples*

We obtained approximation formulas for the Hadamard fractional integrals. The error caused by such decompositions is given later in Section 7.3. In this section we study several cases, comparing the solution with the approximations. To gather more information on the accuracy, we evaluate the error using the distance

$$E = \sqrt{\int_a^b \left({}_a\mathcal{I}_t^\alpha x(t) - {}_a\tilde{\mathcal{I}}_t^\alpha x(t) \right)^2 dt},$$

where $_a\tilde{\mathcal{I}}_t^\alpha x(t)$ is the approximated value.

To begin with, we consider $\alpha = 0.5$ and functions $x_1(t) = \ln t$ and $x_2(t) = 1$ with $t \in [1, 10]$. Then,

$$_1\mathcal{I}_t^{0.5} x_1(t) = \frac{\sqrt{\ln^3 t}}{\Gamma(2.5)} \text{ and } _1\mathcal{I}_t^{0.5} x_2(t) = \frac{\sqrt{\ln t}}{\Gamma(1.5)}$$

(cf. (Kilbas, Srivastava and Trujillo, 2006, Property 2.24)). We consider the expansion formula for $n = 2$ as in (7.17) for both cases. We obtain then the approximations

$$_1\mathcal{I}_t^{0.5}x_1(t) \approx \left[A_0(0.5, N) + A_1(0.5, N) + \sum_{p=3}^{N} B(0.5, p)\frac{p-2}{p-1} \right] \sqrt{\ln^3 t}$$

and

$$_1\mathcal{I}_t^{0.5}x_2(t) \approx \left[A_0(0.5, N) + \sum_{p=3}^{N} B(0.5, p) \right] \sqrt{\ln t}.$$

The results are exemplified in Figs. 7.7(a) and 7.7(b). As can be seen, the value $N = 3$ is enough in order to obtain a good accuracy in the sense of the error function.

We now test the approximation on the power functions $x_3(t) = t^4$ and $x_4(t) = t^9$, with $t \in [1, 2]$. Observe first that

$$_1\mathcal{I}_t^{0.5}(t^k) = \frac{1}{\Gamma(0.5)} \int_1^t \left(\ln \frac{t}{\tau} \right)^{-0.5} \tau^{k-1} d\tau = \frac{t^k}{\Gamma(0.5)} \int_0^{\ln t} \xi^{-0.5} e^{-\xi k} d\xi$$

by the change of variables $\xi = \ln \frac{t}{\tau}$. In our cases,

$$_1\mathcal{I}_t^{0.5}(t^4) \approx \frac{0.8862269255}{\Gamma(0.5)} t^4 \mathrm{erf}(2\sqrt{\ln t})$$

and

$$_1\mathcal{I}_t^{0.5}(t^9) \approx \frac{0.5908179503}{\Gamma(0.5)} t^9 \mathrm{erf}(3\sqrt{\ln t}),$$

where erf is the error function. In Figs. 7.8(a) and 7.8(b) we show approximations for several values of N. We mention that, as N increases, the error decreases and thus we obtain a better approximation.

Another way to obtain different expansion formulas is to vary n. To exemplify, we choose the previous test functions x_i, for $i = 1, 2, 3, 4$, and consider the cases $n = 2, 3, 4$ with $N = 5$ fixed. The results are shown in Figs. 7.9(a), 7.9(b), 7.9(c) and 7.9(d). Observe that as n increases, the error may increase. This can be easily explained by analysis of the error formula, and the values of the sequence $x_{(k,0)}$ involved. For example, for x_4 we have $x_{(k,0)}(t) = 9^k t^9$, for $k = 0 \ldots, n$. This suggests that, when we increase the value of n and the function grows fast, in order to obtain a better accuracy on the method, the value of N should also increase.

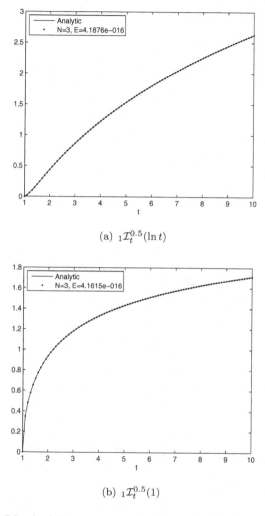

(a) $_1\mathcal{I}_t^{0.5}(\ln t)$

(b) $_1\mathcal{I}_t^{0.5}(1)$

Fig. 7.7   Analytic *versus* numerical approximation for $n = 2$.

7.3   Error analysis

In the previous section we deduced an approximation formula for the left Riemann–Liouville fractional integral (Eq. (7.10)). The order of magnitude of the coefficients that we ignore during this procedure is small for the examples that we have chosen (Tables 7.1 and 7.2). The aim of this section is

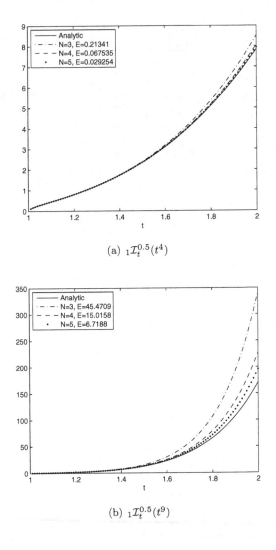

(a) $_1\mathcal{I}_t^{0.5}(t^4)$

(b) $_1\mathcal{I}_t^{0.5}(t^9)$

Fig. 7.8   Analytic *versus* numerical approximation for $n = 2$.

to obtain an estimation for the error, when considering sums up to order N.

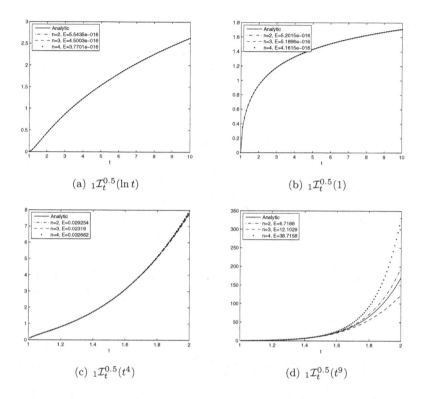

(a) $_1\mathcal{I}_t^{0.5}(\ln t)$

(b) $_1\mathcal{I}_t^{0.5}(1)$

(c) $_1\mathcal{I}_t^{0.5}(t^4)$

(d) $_1\mathcal{I}_t^{0.5}(t^9)$

Fig. 7.9 Analytic *versus* numerical approximation for $n = 2, 3, 4$ and $N = 5$.

We proved that

$$_aI_t^\alpha x(t) = \frac{(t-a)^\alpha}{\Gamma(\alpha+1)}x(a) + \cdots + \frac{(t-a)^{\alpha+n-1}}{\Gamma(\alpha+n)}x^{(n-1)}(a)$$

$$+ \frac{(t-a)^{\alpha+n-1}}{\Gamma(\alpha+n)}\int_a^t \left(1 - \frac{\tau-a}{t-a}\right)^{\alpha+n-1} x^{(n)}(\tau)d\tau.$$

Expanding up to order N the binomial, we get

$$\left(1 - \frac{\tau-a}{t-a}\right)^{\alpha+n-1} = \sum_{p=0}^{N} \frac{\Gamma(p-\alpha-n+1)}{\Gamma(1-\alpha-n)\,p!}\left(\frac{\tau-a}{t-a}\right)^p + R_N(\tau),$$

where

$$R_N(\tau) = \sum_{p=N+1}^{\infty} \frac{\Gamma(p-\alpha-n+1)}{\Gamma(1-\alpha-n)\,p!}\left(\frac{\tau-a}{t-a}\right)^p.$$

Since $\tau \in [a, t]$, we easily deduce an upper bound for $R_N(\tau)$:

$$|R_N(\tau)| \leq \sum_{p=N+1}^{\infty} \left| \frac{\Gamma(p - \alpha - n + 1)}{\Gamma(1 - \alpha - n)\, p!} \right| = \sum_{p=N+1}^{\infty} \left| \binom{\alpha + n - 1}{p} \right|$$

$$\leq \sum_{p=N+1}^{\infty} \frac{e^{(\alpha+n-1)^2 + \alpha + n - 1}}{p^{\alpha+n}}$$

$$\leq \int_{N}^{\infty} \frac{e^{(\alpha+n-1)^2 + \alpha + n - 1}}{p^{\alpha+n}}\, dp = \frac{e^{(\alpha+n-1)^2 + \alpha + n - 1}}{(\alpha + n - 1) N^{\alpha+n-1}}.$$

Thus, we obtain an estimation for the truncation error E_{tr}:

$$|E_{tr}(t)| \leq L_n \frac{(t - a)^{\alpha+n} e^{(\alpha+n-1)^2 + \alpha + n - 1}}{\Gamma(\alpha + n)(\alpha + n - 1) N^{\alpha+n-1}},$$

where $L_n = \max\limits_{\tau \in [a,t]} |x^{(n)}(\tau)|$.

We proceed with an estimation for the error on the approximation for the Hadamard fractional integral. We have proven before that

$$_aI_t^\alpha x(t) = \frac{x(a)}{\Gamma(\alpha + 1)} \left(\ln \frac{t}{a} \right)^\alpha + \frac{a\dot{x}(a)}{\Gamma(\alpha + 2)} \left(\ln \frac{t}{a} \right)^{\alpha+1}$$

$$+ \frac{(a\dot{x}(a) + a^2 \ddot{x}(a))}{\Gamma(\alpha + 3)} \left(\ln \frac{t}{a} \right)^{\alpha+2}$$

$$+ \frac{\left(\ln \frac{t}{a} \right)^{\alpha+2}}{\Gamma(\alpha + 3)} \int_a^t \sum_{p=0}^{\infty} \Gamma_1(\alpha, p, t)(\dot{x}(\tau) + 3\tau\ddot{x}(\tau) + \tau^2\dddot{x}(\tau))d\tau.$$

When we consider finite sums up to order N, the error is given by

$$|E_{tr}(t)| = \left| \frac{1}{\Gamma(\alpha + 3)} \left(\ln \frac{t}{a} \right)^{\alpha+2} \int_a^t R_N(\tau)(\dot{x}(\tau) + 3\tau\ddot{x}(\tau) + \tau^2\dddot{x}(\tau))d\tau \right|$$

with

$$R_N(\tau) = \sum_{p=N+1}^{\infty} \frac{\Gamma(p - \alpha - 2)}{\Gamma(-\alpha - 2)p!} \frac{\left(\ln \frac{\tau}{a} \right)^p}{\left(\ln \frac{t}{a} \right)^p}.$$

Since $\tau \in [a, t]$, we have

$$|R_N(\tau)| \leq \sum_{p=N+1}^{\infty} \left| \binom{\alpha + 2}{p} \right| \leq \sum_{p=N+1}^{\infty} \frac{e^{(\alpha+2)^2 + \alpha + 2}}{p^{\alpha+3}}$$

$$\leq \int_{N}^{\infty} \frac{e^{(\alpha+2)^2 + \alpha + 2}}{p^{\alpha+3}}\, dp = \frac{e^{(\alpha+2)^2 + \alpha + 2}}{(\alpha + 2) N^{\alpha+2}}.$$

Therefore,

$$|E_{tr}(t)| \leq \frac{e^{(\alpha+2)^2+\alpha+2}}{(\alpha+2)N^{\alpha+2}} \frac{\left(\ln \frac{t}{a}\right)^{\alpha+2}}{\Gamma(\alpha+3)}$$
$$\times \left[(t-a)L_1(t) + 3(t-a)^2 L_2(t) + (t-a)^3 L_3(t)\right],$$

where

$$L_i(t) = \max_{\tau \in [a,t]} |x^{(i)}(\tau)|, \quad i \in \{1,2,3\}.$$

We remark that the error formula tends to zero as N increases. Moreover, if we consider the approximation

$$_a\mathcal{I}_t^\alpha x(t) \approx \sum_{i=0}^{n} A_i(\alpha, N) \left(\ln \frac{t}{a}\right)^{\alpha+i} x_{i,0}(t)$$
$$+ \sum_{p=n+1}^{N} B(\alpha, p) \left(\ln \frac{t}{a}\right)^{\alpha+n-p} V_p(t)$$

with $N \geq n+1$ and

$$A_i(\alpha, N) = \frac{1}{\Gamma(\alpha+i+1)} \left[1 + \sum_{p=n-i+1}^{N} \frac{\Gamma(p-\alpha-n)}{\Gamma(-\alpha-i)(p-n+i)!}\right],$$

then the error is bounded by the expression

$$|E_{tr}(t)| \leq L_n(t) \frac{e^{(\alpha+n)^2+\alpha+n}}{\Gamma(\alpha+n+1)(\alpha+n)N^{\alpha+n}} \left(\ln \frac{t}{a}\right)^{\alpha+n} (t-a),$$

where

$$L_n(t) = \max_{\tau \in [a,t]} |x_{n,1}(\tau)|.$$

For the right Hadamard integral, the error is bounded by

$$|E_{tr}(t)| \leq L_n(t) \frac{e^{(\alpha+n)^2+\alpha+n}}{\Gamma(\alpha+n+1)(\alpha+n)N^{\alpha+n}} \left(\ln \frac{b}{t}\right)^{\alpha+n} (b-t),$$

where

$$L_n(t) = \max_{\tau \in [t,b]} |x_{n,1}(\tau)|.$$

Chapter 8

Direct methods

In the presence of fractional operators, the same ideas that were discussed in Section 2.1.3, are applied to discretize the problem. Many works can be found in the literature that use different types of basis functions to establish Ritz-like methods for the fractional calculus of variations and optimal control. Nevertheless, finite differences have received less interest. A brief introduction of using finite differences has been made in (Riewe, 1996), which can be regarded as a predecessor to what we call here an Euler-like direct method. A generalization of Leitmann's direct method can be found in (Almeida and Torres, 2010), while (Lotfi, Dehghan and Yousefi, 2011) discusses the Ritz direct method for optimal control problems that can easily be reduced to a problem of the calculus of variations.

8.1 Finite differences for fractional derivatives

Recall the definitions of Grünwald–Letnikov, e.g. (3.4). It exhibits a finite difference nature involving an infinite series. For numerical purposes we need a finite sum in (3.4). Given a grid on $[a, b]$ as $a = t_0, t_1, \ldots, t_n = b$, where $t_i = t_0 + ih$ for some $h > 0$, $i = 0, \ldots, n$, we approximate the left Riemann–Liouville derivative as

$$_aD_t^\alpha x(t_i) \approx \frac{1}{h^\alpha} \sum_{k=0}^{i} (\omega_k^\alpha) x(t_i - kh), \qquad (8.1)$$

where

$$(\omega_k^\alpha) = (-1)^k \binom{\alpha}{k} = \frac{\Gamma(k - \alpha)}{\Gamma(-\alpha)\Gamma(k + 1)}.$$

Remark 8.1. Similarly, one can approximate the right Riemann–Liouville derivative by

$$_tD_b^\alpha x(t_i) \approx \frac{1}{h^\alpha} \sum_{k=0}^{n-i} (\omega_k^\alpha)\, x(t_i + kh). \qquad (8.2)$$

Remark 8.2. The Grünwald–Letnikov approximation of Riemann–Liouville is a first order approximation (Podlubny, 1999), i.e.,

$$_aD_t^\alpha x(t_i) = \frac{1}{h^\alpha} \sum_{k=0}^{i} (\omega_k^\alpha)\, x(t_i - kh) + \mathcal{O}(h).$$

Remark 8.3. It has been shown that the implicit Euler method solution to a certain fractional partial differential equation based on Grünwald–Letnikov approximation to the fractional derivative is unstable (Meerschaert and Tadjeran, 2004). Therefore, discretizing fractional derivatives and shifted Grünwald–Letnikov derivatives are used, and despite the slight difference they exhibit a stable performance, at least for certain cases. The left shifted Grünwald–Letnikov derivative is defined by

$$_a^{sGL}D_t^\alpha x(t_i) \approx \frac{1}{h^\alpha} \sum_{k=0}^{i} (\omega_k^\alpha)\, x(t_i - (k-1)h).$$

Other finite difference approximations can be found in the literature. Specifically, we refer to (Diethelm *et al.*, 2005), Diethelm's backward finite differences formula for the Caputo fractional derivative, with $0 < \alpha < 2$ and $\alpha \neq 1$, that is an approximation of order $\mathcal{O}(h^{2-\alpha})$:

$$_a^C D_t^\alpha x(t_i) \approx \frac{h^{-\alpha}}{\Gamma(2-\alpha)} \sum_{j=0}^{i} a_{i,j} \left(x_{i-j} - \sum_{k=0}^{\lfloor \alpha \rfloor} \frac{(i-j)^k h^k}{k!} x^{(k)}(a) \right),$$

where

$$a_{i,j} = \begin{cases} 1, & \text{if } i = 0, \\ (j+1)^{1-\alpha} - 2j^{1-\alpha} + (j-1)^{1-\alpha}, & \text{if } 0 < j < i, \\ (1-\alpha)i^{-\alpha} - i^{1-\alpha} + (i-1)^{1-\alpha}, & \text{if } j = i. \end{cases}$$

8.2   Euler-like direct method for variational problems

8.2.1   *Euler's classic direct method*

Euler's method in the classical theory of the calculus of variations uses finite difference approximations for derivatives and is also referred to as the method of finite differences. The basic idea of this method is that instead of considering the values of a functional

$$J[x] = \int_a^b L(t, x(t), \dot{x}(t)) dt$$

with boundary conditions $x(a) = x_a$ and $x(b) = x_b$, on arbitrary admissible curves, we only track the values at an $n + 1$ grid points, t_i, $i = 0, \ldots, n$, of the interested time interval (Pooseh, Almeida and Torres, 2012b). The functional $J[x]$ is then transformed into a function $\Psi(x(t_1), x(t_2), \ldots, x(t_{n-1}))$ of the values of the unknown function on mesh points. Assuming $h = t_i - t_{i-1}$, $x(t_i) = x_i$ and $\dot{x}_i \approx \frac{x_i - x_{i-1}}{h}$, one has

$$J[x] \approx \Psi(x_1, x_2, \ldots, x_{n-1}) = h \sum_{i=1}^{n} L\left(t_i, x_i, \frac{x_i - x_{i-1}}{h}\right),$$

$$x_0 = x_a, \quad x_n = x_b.$$

The desired values of x_i, $i = 1, \ldots, n-1$, are the extremizers of the multi-variable function Ψ, which are solution to the system

$$\frac{\partial \Psi}{\partial x_i} = 0, \quad i = 1, \ldots, n-1.$$

The fact that only two terms in the sum, $(i-1)$th and ith, depend on x_i, makes it rather easy to find the extremals of Ψ solving a system of algebraic equations. For each n, we obtain a polygonal line which is an approximate solution of the original problem. It has been shown that passing to the limit as $h \to 0$, the linear system corresponding to finding the extremals of Ψ is equivalent to the Euler–Lagrange equation for the problem (Tuckey, 1993).

8.2.2 *Euler-like direct method*

As mentioned earlier, we consider a simple version of fractional variational problems where the fractional term has a Riemann–Liouville derivative on a finite time interval $[a, b]$. The boundary conditions are given and we approximate the derivative using Grünwald–Letnikov approximation given by (8.1). In this context, we discretize the functional in (4.1) using a simple quadrature rule on the mesh points, $a = t_0, t_1, , \ldots, t_n = b$, with $h = \frac{b-a}{n}$. The goal is to find the values $x_1, \ldots, x_{n-1}$ of the unknown function x at the points t_i, $i = 1, \ldots, n-1$. The values of x_0 and x_n are given. Applying the quadrature rule gives

$$J[x] = \sum_{i=1}^{n} \int_{t_{i-1}}^{t_i} L(t_i, x_i, {}_aD_{t_i}^\alpha x_i) dt \approx \sum_{i=1}^{n} h L(t_i, x_i, {}_aD_{t_i}^\alpha x_i),$$

and by approximating the fractional derivatives at mesh points using (8.1) we have

$$J[x] \approx \sum_{i=1}^{n} hL\left(t_i, x_i, \frac{1}{h^\alpha} \sum_{k=0}^{i} (\omega_k^\alpha) \, x_{i-k}\right). \tag{8.3}$$

Hereafter the procedure is the same as in the classical case. The right-hand side of (8.3) can be regarded as a function Ψ of $n-1$ unknowns $\mathbf{x} = (x_1, x_2, \ldots, x_{n-1})$,

$$\Psi(\mathbf{x}) = \sum_{i=1}^{n} hL\left(t_i, x_i, \frac{1}{h^\alpha} \sum_{k=0}^{i} (\omega_k^\alpha) \, x_{i-k}\right). \tag{8.4}$$

To find an extremizer for Ψ, one solves the following system of algebraic equations:

$$\frac{\partial \Psi}{\partial x_i} = 0, \qquad i = 1, \ldots, n-1. \tag{8.5}$$

Unlike the classical case, all terms, starting from ith term, in (8.4) depend on x_i and we have

$$\frac{\partial \Psi}{\partial x_i} = h \frac{\partial L}{\partial x}(t_i, x_i, {}_aD_{t_i}^\alpha x_i) + h \sum_{k=0}^{n-i} \frac{1}{h^\alpha} (\omega_k^\alpha) \frac{\partial L}{\partial \, {}_aD_t^\alpha x}(t_{i+k}, x_{i+k}, {}_aD_{t_{i+k}}^\alpha x_{i+k}). \tag{8.6}$$

Equating the right-hand side of (8.6) with zero, one has

$$\frac{\partial L}{\partial x}(t_i, x_i, {}_aD_{t_i}^\alpha x_i) + \frac{1}{h^\alpha} \sum_{k=0}^{n-i} (\omega_k^\alpha) \frac{\partial L}{\partial \, {}_aD_t^\alpha x}(t_{i+k}, x_{i+k}, {}_aD_{t_{i+k}}^\alpha x_{i+k}) = 0.$$

Passing to the limit and considering the approximation formula for the right Riemann–Liouville derivative, Eq. (8.2), it is straightforward to verify that:

Theorem 8.1. *The Euler-like method for a fractional variational problem of the form* (4.1) *is equivalent to the fractional Euler–Lagrange equation*

$$\frac{\partial L}{\partial x} + {}_tD_b^\alpha \frac{\partial L}{\partial \, {}_aD_t^\alpha x} = 0,$$

as the mesh size, h, tends to zero.

Proof. Consider a minimizer $(x_1, \ldots, x_{n-1})$ of Ψ, a variation function $\eta \in C[a, b]$ with $\eta(a) = \eta(b) = 0$ and define $\eta_i = \eta(t_i)$, for $i = 0, \ldots, n$. We remark that $\eta_0 = \eta_n = 0$ and that $(x_1 + \epsilon\eta_1, \ldots, x_{n-1} + \epsilon\eta_{n-1})$ is a variation of $(x_1, \ldots, x_{n-1})$, with $|\epsilon| < r$, for some fixed $r > 0$. Therefore, since

$(x_1, \ldots, x_{n-1})$ is a minimizer for Ψ, proceeding with Taylor's expansion, we deduce

$$0 \leq \Psi(x_1 + \epsilon\eta_1, \ldots, x_{n-1} + \epsilon\eta_{n-1}) - \Psi(x_1, \ldots, x_{n-1})$$

$$= \epsilon \sum_{i=1}^{n} h \left[\frac{\partial L}{\partial x}[i]\eta_i + \frac{\partial L}{\partial_a D_t^\alpha}[i] \frac{1}{h^\alpha} \sum_{k=0}^{i} (\omega_k^\alpha)\eta_{i-k} \right] + \mathcal{O}(\epsilon),$$

where

$$[i] = \left(t_i, x_i, \frac{1}{h^\alpha} \sum_{k=0}^{i} (\omega_k^\alpha) x_{i-k} \right).$$

Since ϵ takes any value, it follows that

$$\sum_{i=1}^{n} h \left[\frac{\partial L}{\partial x}[i]\eta_i + \frac{\partial L}{\partial_a D_t^\alpha}[i] \frac{1}{h^\alpha} \sum_{k=0}^{i} (\omega_k^\alpha)\eta_{i-k} \right] = 0. \qquad (8.7)$$

On the other hand, since $\eta_0 = 0$, reordering the terms of the sum, it follows immediately that

$$\sum_{i=1}^{n} \frac{\partial L}{\partial_a D_t^\alpha}[i] \sum_{k=0}^{i} (\omega_k^\alpha)\eta_{i-k} = \sum_{i=1}^{n} \eta_i \sum_{k=0}^{n-i} (\omega_k^\alpha) \frac{\partial L}{\partial_a D_t^\alpha}[i+k].$$

Substituting this relation into Eq. (8.7), we obtain

$$\sum_{i=1}^{n} \eta_i h \left[\frac{\partial L}{\partial x}[i] + \frac{1}{h^\alpha} \sum_{k=0}^{n-i} (\omega_k^\alpha) \frac{\partial L}{\partial_a D_t^\alpha}[i+k] \right] = 0.$$

Since η_i is arbitrary, for $i = 1, \ldots, n-1$, we deduce that

$$\frac{\partial L}{\partial x}[i] + \frac{1}{h^\alpha} \sum_{k=0}^{n-i} (\omega_k^\alpha) \frac{\partial L}{\partial_a D_t^\alpha}[i+k] = 0, \quad \text{for } i = 1, \ldots, n-1.$$

Let us study the case when n goes to infinity. Let $\bar{t} \in]a, b[$ and $i \in \{1, \ldots, n\}$ such that $t_{i-1} < \bar{t} \leq t_i$. First observe that in such case, we also have $i \to \infty$ and $n - i \to \infty$. In fact, let $i \in \{1, \ldots, n\}$ be such that

$$a + (i-1)h < \bar{t} \leq a + ih.$$

So, $i < (\bar{t} - a)/h + 1$, which implies that

$$n - i > n \frac{b - \bar{t}}{b - a} - 1.$$

Then

$$\lim_{n \to \infty, i \to \infty} t_i = \bar{t}.$$

Assume that there exists a function $\bar{x} \in C[a, b]$ satisfying

$$\forall \epsilon > 0 \, \exists N \, \forall n \geq N \, : \, |x_i - \bar{x}(t_i)| < \epsilon, \quad \forall i = 1, \ldots, n - 1.$$

As $\bar{x}$ is uniformly continuous, we have

$$\forall \epsilon > 0 \, \exists N \, \forall n \geq N \, : \, |x_i - \bar{x}(\bar{t})| < \epsilon, \quad \forall i = 1, \ldots, n - 1.$$

By the continuity assumption of $\bar{x}$, we deduce that

$$\lim_{n \to \infty, i \to \infty} \frac{1}{h^\alpha} \sum_{k=0}^{n-i} (\omega_k^\alpha) \frac{\partial L}{\partial_a D_t^\alpha}[i + k] = {_tD_b^\alpha} \frac{\partial L}{\partial_a D_t^\alpha}(\bar{t}, \bar{x}(\bar{t}), {_aD_t^\alpha}\bar{x}(\bar{t})).$$

For n sufficiently large (and therefore i also sufficiently large),

$$\lim_{n \to \infty, i \to \infty} \frac{\partial L}{\partial x}[i] = \frac{\partial L}{\partial x}(\bar{t}, \bar{x}(\bar{t}), {_aD_t^\alpha}\bar{x}(\bar{t})).$$

In conclusion,

$$\frac{\partial L}{\partial x}(\bar{t}, \bar{x}(\bar{t}), {_aD_t^\alpha}\bar{x}(\bar{t})) + {_tD_b^\alpha} \frac{\partial L}{\partial_a D_t^\alpha}(\bar{t}, \bar{x}(\bar{t}), {_aD_t^\alpha}\bar{x}(\bar{t})) = 0. \qquad (8.8)$$

Using the continuity condition, we prove that the fractional Euler–Lagrange equation (8.8) for all values on the closed interval $a \leq t \leq b$ holds. $\qquad \square$

8.2.3   *Examples*

Now we apply the Euler-like direct method to some test problems for which the exact solutions are in hand. Although we propose problems on to the interval $[0, 1]$, moving to arbitrary intervals is a matter of more computations. To measure the errors related to approximations, different norms can be used. Since a direct method seeks for the function values at certain points, we use the maximum norm to determine how close we can get to the exact value at that point. Assume that the exact value of the function x, at the point t_i, is $x(t_i)$ and it is approximated by x_i. The error is defined as

$$E = \max\{|x(t_i) - x_i|, \ i = 1, 2, \ldots, n\}. \qquad (8.9)$$

Example 8.1. Our goal here is to minimize a quadratic Lagrangian on $[0, 1]$ with fixed boundary conditions. Consider the following minimization problem:

$$J[x] = \int_0^1 \left({_0D_t^{0.5}}x(t) - \frac{2}{\Gamma(2.5)} t^{1.5} \right)^2 dt \longrightarrow \min \qquad (8.10)$$

$$x(0) = 0, \quad x(1) = 1.$$

Since the Lagrangian is always positive, problem (8.10) attains its minimum when

$$_0D_t^{0.5}x(t) - \frac{2}{\Gamma(2.5)}t^{1.5} = 0,$$

and has the obvious solution of the form $x(t) = t^2$ because $_0D_t^{0.5}t^2 = \frac{2}{\Gamma(2.5)}t^{1.5}$.

To begin with, we approximate the fractional derivative by

$$_0D_t^{0.5}x(t_i) \approx \frac{1}{h^{0.5}}\sum_{k=0}^{i}\left(\omega_k^{0.5}\right)x(t_i - kh)$$

for a fixed $h > 0$. The functional is now transformed into

$$J[x] \approx \int_0^1 \left(\frac{1}{h^{0.5}}\sum_{k=0}^{i}\left(\omega_k^{0.5}\right)x_{i-k} - \frac{2}{\Gamma(2.5)}t^{1.5}\right)^2 dt.$$

Finally, we approximate the integral by a rectangular rule and end with the discrete problem

$$\Psi(\mathbf{x}) = \sum_{i=1}^{n}h\left(\frac{1}{h^{0.5}}\sum_{k=0}^{i}\left(\omega_k^{0.5}\right)x_{i-k} - \frac{2}{\Gamma(2.5)}t_i^{1.5}\right)^2.$$

Since the Lagrangian in this example is quadratic, system (8.5) has a linear form and therefore is easy to solve. Other problems may end with a system of nonlinear equations. Simple calculations lead to the system

$$\mathbf{Ax} = \mathbf{b}, \tag{8.11}$$

in which

$$\mathbf{A} = \begin{bmatrix} \sum_{i=0}^{n-1}A_i^2 & \sum_{i=1}^{n-1}A_iA_{i-1} & \cdots & \sum_{i=n-2}^{n-1}A_iA_{i-(n-2)} \\ \sum_{i=0}^{n-2}A_iA_{i+1} & \sum_{i=1}^{n-2}A_i^2 & \cdots & \sum_{i=n-3}^{n-2}A_iA_{i-(n-3)} \\ \sum_{i=0}^{n-3}A_iA_{i+2} & \sum_{i=1}^{n-3}A_iA_{i+1} & \cdots & \sum_{i=n-4}^{n-3}A_iA_{i-(n-4)} \\ \vdots & \vdots & \ddots & \vdots \\ \sum_{i=0}^{1}A_iA_{i+n-2} & \sum_{i=0}^{1}A_iA_{i+n-3} & \cdots & \sum_{i=0}^{1}A_i^2 \end{bmatrix}$$

where $A_i = (-1)^i h^{1.5}\binom{0.5}{i}$, $\mathbf{x} = (x_1, \cdots, x_{n-1})^T$ and $\mathbf{b} = (b_1, \cdots, b_{n-1})^T$ with

$$b_i = \sum_{k=0}^{n-i}\frac{2h^2 A_k}{\Gamma(2.5)}t_{k+i}^{1.5} - A_{n-i}A_0 - \left(\sum_{k=0}^{n-i}A_kA_{k+i}\right).$$

The linear system (8.11) is easily solved for different values of n (see Appendix). As indicated in Fig. 8.1, by increasing the value of n we get better solutions.

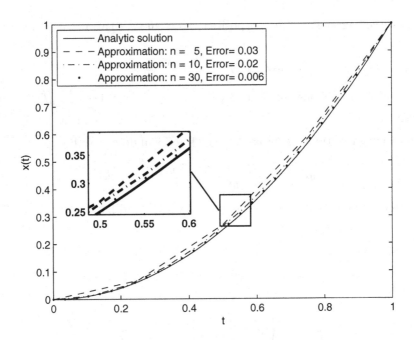

Fig. 8.1   Analytic and approximate solutions of Example 8.1.

Let us now move to another example for which the solution is obtained by the fractional Euler–Lagrange equation.

Example 8.2. Consider the following minimization problem:

$$\begin{cases} J[x] = \int_0^1 \left({}_0D_t^{0.5}x(t) - \dot{x}^2(t) \right) dt \to \min \\ x(0) = 0, \; x(1) = 1. \end{cases} \tag{8.12}$$

In this case the only way to get a solution is by use of Euler–Lagrange equations. The Lagrangian depends not only on the fractional derivative, but also on the first-order derivative of the function. The Euler–Lagrange equation for this setting becomes

$$\frac{\partial L}{\partial x} + {}_tD_b^\alpha \frac{\partial L}{\partial {}_aD_t^\alpha} - \frac{d}{dt}\left(\frac{\partial L}{\partial \dot{x}} \right) = 0$$

and, by direct computations, a necessary condition for x to be a minimizer of (8.12) is

$${}_tD_1^\alpha 1 + 2\ddot{x}(t) - 0, \text{ or } \ddot{x}(t) = \frac{1}{2\Gamma(1-\alpha)}(1-t)^{-\alpha}.$$

Subject to the given boundary conditions, the above second-order ordinary differential equation has the solution

$$x(t) = -\frac{1}{2\Gamma(3-\alpha)}(1-t)^{2-\alpha} + \left(1 - \frac{1}{2\Gamma(3-\alpha)}\right)t + \frac{1}{2\Gamma(3-\alpha)}. \quad (8.13)$$

Discretizing problem (8.12) with the same assumptions of Example 8.1 ends in a linear system of the form

$$\begin{bmatrix} 2 & -1 & 0 & 0 & \cdots & 0 & 0 \\ -1 & 2 & -1 & 0 & \cdots & 0 & 0 \\ 0 & -1 & 2 & -1 & \cdots & 0 & 0 \\ \vdots & \vdots & \vdots & \vdots & \ddots & \vdots & \vdots \\ 0 & 0 & 0 & 0 & \cdots & -1 & 2 \end{bmatrix} \begin{bmatrix} x_1 \\ x_2 \\ x_3 \\ \vdots \\ x_{n-1} \end{bmatrix} = \begin{bmatrix} b_1 \\ b_2 \\ b_3 \\ \vdots \\ b_{n-1} \end{bmatrix}, \quad (8.14)$$

where

$$b_i = \frac{h}{2} \sum_{k=0}^{n-i-1} (-1)^k h^{0.5} \binom{0.5}{k}, \qquad i = 1, 2, \cdots, n-2,$$

and

$$b_{n-1} = \frac{h}{2} \sum_{k=0}^{1} \left((-1)^k h^{0.5} \binom{0.5}{k}\right) + x_n.$$

System (8.14) is linear and can be solved for any n to reach the desired accuracy (see Appendix). The analytic solution together with some approximated solutions are shown in Fig. 8.2.

Both examples above end with linear systems and their solvability is simply dependant on the matrix of coefficients. Now we try our method on a more complicated problem, yet analytically solvable with an oscillating solution.

Example 8.3. Let $0 < \alpha < 1$ and we are supposed to minimize a functional with the following Lagrangian on $[0, 1]$:

$$L = \left(_0D_t^{0.5}x(t) - \frac{16\Gamma(6)}{\Gamma(5.5)}t^{4.5} + \frac{20\Gamma(4)}{\Gamma(3.5)}t^{2.5} - \frac{5}{\Gamma(1.5)}t^{0.5}\right)^4.$$

This example has an obvious solution too. Since L is positive, $\int_0^1 L\,dt$ subject to the boundary conditions $x(0) = 0$ and $x(1) = 1$ has a minimizer of the form

$$x(t) = 16t^5 - 20t^3 + 5t.$$

Note that

$$_aD_t^\alpha(t-a)^\nu = \frac{\Gamma(\nu+1)}{\Gamma(\nu+\alpha)}(t-a)^{\nu-\alpha}.$$

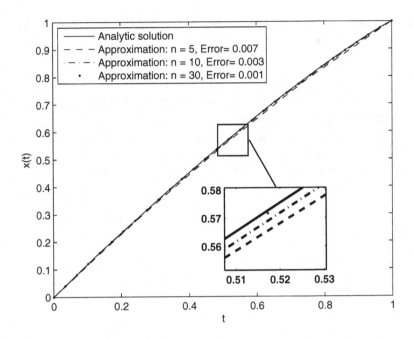

Fig. 8.2   Analytic and approximate solutions of Example 8.2.

The appearance of a fourth power in the Lagrangian, results in a nonlinear system as we apply the Euler-like direct method to this problem. For $j = 1, \ldots, n - 1$, we have

$$\sum_{i=j}^{n} \left(\omega_{i-j}^{0.5}\right) \left(\frac{1}{h^{0.5}} \sum_{k=0}^{i} \left(\omega_k^{0.5}\right) x_{i-k} - \phi(t_i)\right)^3 = 0, \qquad (8.15)$$

where

$$\phi(t) = \frac{16\Gamma(6)}{\Gamma(5.5)} t^{4.5} + \frac{20\Gamma(4)}{\Gamma(3.5)} t^{2.5} - \frac{5}{\Gamma(1.5)} t^{0.5}.$$

System (8.15) is solved for different values of n (see Appendix) and the results are depicted in Fig. 8.3.

These examples show that an Euler-like direct method reduces a variational problem to a system of algebraic equations. When the resulting system is linear, better solutions are obtained by increasing the number of mesh points as long as the resulted matrix of coefficients is invertible. The method is very fast in this case.

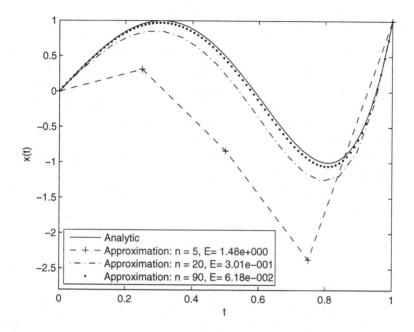

Fig. 8.3   Analytic and approximate solutions of Example 8.3.

The situation is completely different when the problem ends with a nonlinear system. Table 8.1 summarizes the results regarding the running time and the error.

Table 8.1   Number of mesh points, n, with corresponding running time in seconds, T, and error, E (8.9).

	n	T	E
Example 1	5	0.00019668	0.0264
	10	0.00028297	0.0158
	30	0.00098318	0.0065
Example 2	5	0.00024053	0.0070
	10	0.00030209	0.0035
	30	0.00073457	0.0012
Example 3	5	0.0126	1.4787
	20	0.2012	0.3006
	90	26.355	0.0618

8.3   A discrete-time method on the first variation

The fact that the first variation of a variational functional must vanish along
an extremizer is the base of most effective solution schemes to solve problems of the calculus of variations. We generalize the method to variational
problems involving fractional-order derivatives. First-order splines are used
as variations, for which fractional derivatives are known. The Grünwald–
Letnikov definition of fractional derivative is used as a consequence of its
intrinsic discrete nature, which leads to straightforward approximations
(Pooseh, Almeida and Torres, 2013e).

The problem under consideration is stated in the following way: find
the extremizers of

$$J[x] = \int_a^b L(t, x(t), {}_aD_t^\alpha x(t)) \, dt \tag{8.16}$$

subject to given boundary conditions $x(a) = x_a$ and $x(b) = x_b$. Here,
$L : [a, b] \times \mathbb{R}^2 \to \mathbb{R}$ is such that $\frac{\partial L}{\partial x}$ and $\frac{\partial L}{\partial_a D_t^\alpha x}$ exist and are continuous
for all triplets $(t, x(t), {}_aD_t^\alpha x(t))$. If x is a solution to the problem and
$\eta : [a, b] \to \mathbb{R}$ is a variation function, i.e., $\eta(a) = \eta(b) = 0$, then the first
variation of J at x, with the variation η, whatever choice of η is taken, must
vanish:

$$
\begin{aligned}
J'[x, \eta] = \int_a^b \Bigg[& \frac{\partial L}{\partial x}(t, x(t), {}_aD_t^\alpha x(t))\eta(t) \\
& + \frac{\partial L}{\partial_a D_t^\alpha x}(t, x(t), {}_aD_t^\alpha x(t)){}_aD_t^\alpha \eta(t) \Bigg] \, dt = 0.
\end{aligned}
\tag{8.17}
$$

Using an integration by parts formula for fractional derivatives and the
Dubois–Reymond lemma, Riewe (1997) proved that if x is an extremizer of
(8.16), then

$$\frac{\partial L}{\partial x}(t, x(t), {}_aD_t^\alpha x(t)) + {}_tD_b^\alpha \left(\frac{\partial L}{\partial_a D_t^\alpha x} \right)(t, x(t), {}_aD_t^\alpha x(t)) = 0$$

(see also (Agrawal, 2002)). This fractional differential equation is called an
Euler–Lagrange equation. For the state of the art on the subject we refer
the reader to the recent book (Malinowska and Torres, 2012). Here, instead
of solving such Euler–Lagrange equations, we apply a discretization over
time and solve a system of algebraic equations. The procedure has proven
to be a successful tool for classical variational problems (Gregory and Lin,
1993; Gregory and Wang, 1990).

The discretization method is the following. Let $n \in \mathbb{N}$ be a fixed parameter and $h = \frac{b-a}{n}$. If we define $t_i = a + ih$, $x_i = x(t_i)$, and $\eta_i = \eta(t_i)$ for $i = 0, \ldots, n$, the integral (8.17) can be approximated by the sum

$$J'[x, \eta)] \approx h \sum_{i=1}^{n} \left[\frac{\partial L}{\partial x}(t_i, x(t_i), {}_aD_{t_i}^{\alpha}x(t_i))\eta(t_i) \right.$$

$$\left. + \frac{\partial L}{\partial\, {}_aD_t^{\alpha}x}(t_i, x(t_i), {}_aD_{t_i}^{\alpha}x(t_i)) {}_aD_{t_i}^{\alpha}\eta(t_i) \right].$$

To compute the fractional derivative, we replace it by the sum as in (8.1), and to find an approximation for x on mesh points one must solve the equation

$$\sum_{i=1}^{n} \left[\frac{\partial L}{\partial x}\left(t_i, x_i, \frac{1}{h^{\alpha}}\sum_{k=0}^{i}(\omega_k^{\alpha})\, x_{i-k}\right) \eta_i \right.$$

$$\left. + \frac{\partial L}{\partial\, {}_aD_t^{\alpha}x}\left(t_i, x_i, \frac{1}{h^{\alpha}}\sum_{k=0}^{i}(\omega_k^{\alpha})\, x_{i-k}\right)\frac{1}{h^{\alpha}}\sum_{k=0}^{i}(\omega_k^{\alpha})\, \eta_{i-k} \right] = 0. \tag{8.18}$$

For different choices of η, one obtains different equations. Here we use simple variations. More precisely, we use first-order splines as the set of variation functions:

$$\eta_j(t) = \begin{cases} \dfrac{t - t_{j-1}}{h} & \text{if } t_{j-1} \le t < t_j, \\ \dfrac{t_{j+1} - t}{h} & \text{if } t_j \le t < t_{j+1}, \\ 0 & \text{otherwise}, \end{cases} \tag{8.19}$$

for $j = 1, \ldots, n - 1$. We remark that conditions $\eta_j(a) = \eta_j(b) = 0$ are fulfilled for all j, and that $\eta_j(t_i) = 0$ for $i \ne j$ and $\eta_j(t_j) = 1$. The fractional derivative of η_j at any point t_i is also computed using approximation (8.1):

$${}_aD_{t_i}^{\alpha}\eta_j(t_i) = \begin{cases} \dfrac{1}{h^{\alpha}}(w_{i-j}^{\alpha}) & \text{if } j \le i, \\ 0 & \text{otherwise}. \end{cases}$$

Using η_j, $j = 1, \ldots, n-1$, and Eq. (8.18), we establish the following system of $n - 1$ algebraic equations with $n - 1$ unknown variables $x_1, \ldots, x_{n-1}$:

$$\begin{cases} \dfrac{\partial L}{\partial x}\{x_1\} + \dfrac{1}{h^{\alpha}}\sum_{i=1}^{n}\left[\dfrac{\partial L}{\partial\, {}_aD_t^{\alpha}x}\{x_i\}(w_{i-1}^{\alpha})\right] = 0, \\[2mm] \dfrac{\partial L}{\partial x}\{x_2\} + \dfrac{1}{h^{\alpha}}\sum_{i=2}^{n}\left[\dfrac{\partial L}{\partial\, {}_aD_t^{\alpha}x}\{x_i\}(w_{i-2}^{\alpha})\right] = 0, \\[2mm] \vdots \\[2mm] \dfrac{\partial L}{\partial x}\{x_{n-1}\} + \dfrac{1}{h^{\alpha}}\sum_{i=n-1}^{n}\left[\dfrac{\partial L}{\partial\, {}_aD_t^{\alpha}x}\{x_i\}(w_{i-n+1}^{\alpha})\right] = 0, \end{cases} \tag{8.20}$$

where we define

$$\{x_i\} = \left(t_i, x_i, \frac{1}{h^\alpha} \sum_{k=0}^{i} (\omega_k^\alpha) x_{i-k}\right).$$

The solution to (8.20), if exists, gives an approximation to the values of the unknown function x on mesh points t_i.

We have considered so far the so-called fundamental or basic problem of the fractional calculus of variations (Malinowska and Torres, 2012). However, other types of problems can be solved applying similar techniques. Let us show how to solve numerically the isoperimetric problem, that is, when in the initial problem the set of admissible functions must satisfy some integral constraint that involves a fractional derivative. We state the fractional isoperimetric problem as follows.

Assume that the set of admissible functions are subject not only to some prescribed boundary conditions, but to some integral constraint, say

$$\int_a^b g(t, x(t), {}_aD_t^\alpha x(t))\, dt = K,$$

for a fixed $K \in \mathbb{R}$. As usual, we assume that $g : [a, b] \times \mathbb{R}^2 \to \mathbb{R}$ is such that $\frac{\partial g}{\partial x}$ and $\frac{\partial g}{\partial_a D_t^\alpha x}$ exist and are continuous. The common procedure to solve this problem follows some simple steps: first we consider the auxiliary function

$$F = \lambda_0 L(t, x(t), {}_aD_t^\alpha x(t)) + \lambda g(t, x(t), {}_aD_t^\alpha x(t)), \tag{8.21}$$

for some constants λ_0 and λ to be determined later. Next, it can be proven that F satisfies the fractional Euler–Lagrange equation and that in case the extremizer does not satisfies the Euler–Lagrange associated to g, then we can take $\lambda_0 = 1$ (cf. (Almeida, Ferreira and Torres, 2012)). In conclusion, the first variation of F evaluated along an extremal must vanish, and so we obtain a system similar to (8.20), replacing L by F. Also, from the integral constraint, we obtain another equation derived by discretization that is used to obtain λ:

$$h \sum_{i=1}^{n} g\left(t_i, x_i, \frac{1}{h^\alpha} \sum_{k=0}^{i} (\omega_k^\alpha) x_{i-k}\right) = K.$$

We show the usefulness of our approximate method with three problems of the fractional calculus of variations.

8.3.1 *Basic fractional variational problems*

Example 8.4. Consider the following variational problem: to minimize the functional

$$J[x] = \int_0^1 \left({}_0D_t^{0.5}x(t) - \frac{2}{\Gamma(2.5)}t^{1.5} \right)^2 dt$$

subject to the boundary conditions $x(0) = 0$ and $x(1) = 1$. It is an easy exercise to verify that the solution is the function $x(t) = t^2$.

We apply our method to this problem, for the variation (8.19). The functional J does not depend on x and is quadratic with respect to the fractional term. Therefore, the first variation is linear. The resulting algebraic system from (8.20) is also linear and easy to solve:

$$\sum_{i=0}^{n-1} \left(\omega_i^{0.5}\right)^2 x_1 + \sum_{i=1}^{n-1} \left(\omega_i^{0.5}\right)\left(\omega_{i-1}^{0.5}\right) x_2 + \sum_{i=2}^{n-1} \left(\omega_i^{0.5}\right)\left(\omega_{i-2}^{0.5}\right) x_3$$

$$+ \cdots + \sum_{i=n-2}^{n-1} \left(\omega_i^{0.5}\right)\left(\omega_{i-(n-2)}^{0.5}\right) x_{n-1}$$

$$= \frac{2h^2}{\Gamma(2.5)} \sum_{i=0}^{n-1} \left(\omega_i^{0.5}\right)(i+1)^{1.5} - \left(\omega_0^{0.5}\right)\left(\omega_{n-1}^{0.5}\right),$$

$$\sum_{i=0}^{n-2} \left(\omega_i^{0.5}\right)\left(\omega_{i+1}^{0.5}\right) x_1 + \sum_{i=0}^{n-2} \left(\omega_i^{0.5}\right)^2 x_2 + \sum_{i=1}^{n-2} \left(\omega_i^{0.5}\right)\left(\omega_{i-1}^{0.5}\right) x_3$$

$$+ \cdots + \sum_{i=n-3}^{n-2} \left(\omega_i^{0.5}\right)\left(\omega_{i-(n-3)}^{0.5}\right) x_{n-1}$$

$$= \frac{2h^2}{\Gamma(2.5)} \sum_{i=0}^{n-2} \left(\omega_i^{0.5}\right)(i+2)^{1.5} - \left(\omega_0^{0.5}\right)\left(\omega_{n-2}^{0.5}\right),$$

$$\vdots$$

$$\sum_{i=0}^{1} \left(\omega_i^{0.5}\right)\left(\omega_{i+n-2}^{0.5}\right) x_1 + \sum_{i=0}^{1} \left(\omega_i^{0.5}\right)\left(\omega_{i+n-3}^{0.5}\right) x_2 + \sum_{i=0}^{1} \left(\omega_i^{0.5}\right)\left(\omega_{i+n-4}^{0.5}\right) x_3$$

$$+ \cdots + \sum_{i=0}^{1} \left(\omega_i^{0.5}\right)^2 x_{n-1}$$

$$= \frac{2h^2}{\Gamma(2.5)} \sum_{i=0}^{1} \left(\omega_i^{0.5}\right)(i+n-1)^{1.5} - \left(\omega_0^{0.5}\right)\left(\omega_1^{0.5}\right).$$

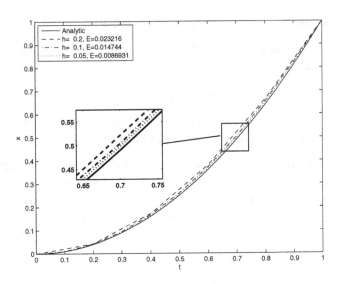

Fig. 8.4   Exact solution *versus* numerical approximations to Example 8.4.

The exact solution together with three numerical approximations, with different discretization step sizes, are depicted in Fig. 8.4.

Example 8.5. Find the minimizer of the functional

$$J[x] = \int_0^1 \left({}_0D_t^{0.5}x(t) - \frac{16\Gamma(6)}{\Gamma(5.5)}t^{4.5} + \frac{20\Gamma(4)}{\Gamma(3.5)}t^{2.5} - \frac{5}{\Gamma(1.5)}t^{0.5} \right)^4 dt$$

subject to $x(0) = 0$ and $x(1) = 1$. The minimum value of this functional is zero and the minimizer is

$$x(t) = 16t^5 - 20t^3 + 5t.$$

Discretizing the first variation as discussed above, leads to a nonlinear system of algebraic equation. Its solution, using different step sizes, is depicted in Fig. 8.5.

8.3.2 *An isoperimetric fractional variational problem*

Example 8.6. Let us search the minimizer of

$$J[x] = \int_0^1 \left(t^4 + \left({}_0D_t^{0.5}x(t) \right)^2 \right) dt$$

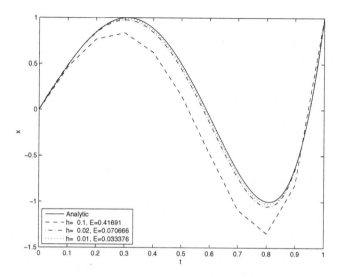

Fig. 8.5   Exact solution *versus* numerical approximations to Example 8.5.

subject to the boundary conditions

$$x(0) = 0 \quad \text{and} \quad x(1) = \frac{16}{15\Gamma(0.5)}$$

and the integral constraint

$$\int_0^1 t^2\, {}_0D_t^{0.5}x(t)\, dt = \frac{1}{5}.$$

In (Almeida and Torres, 2009a) it is shown that the solution to this problem is the function

$$x(t) = \frac{16t^{2.5}}{15\Gamma(0.5)}.$$

As x does not satisfy the fractional Euler–Lagrange equation associated to the integral constraint, one can take $\lambda_0 = 1$ and the auxiliary function (8.21) is $F = t^4 + \left({}_0D_t^{0.5}x(t)\right)^2 + \lambda\, t^2\, {}_0D_t^{0.5}x(t)$. Now we calculate the first variation of $\int_0^1 F\, dt$. An extra unknown, λ, is present in the new setting, which is obtained by discretizing the integral constraint, as explained in Section 8.3. The solutions to the resulting algebraic system, with different step sizes, are given in Fig. 8.6.

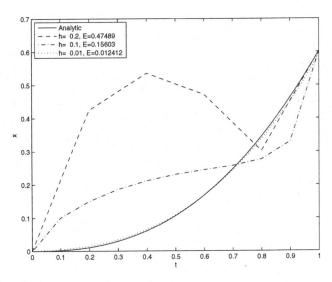

Fig. 8.6   Exact *versus* numerical approximations to the isoperimetric problem of Example 8.6.

Chapter 9

Indirect methods

As in the classical case, indirect methods in fractional sense provide necessary conditions of optimality using the first variation. Fractional Euler–Lagrange equations are now a well-known and well-studied subject in fractional calculus. For a simple problem of the form (4.1), following (Agrawal, 2002), a necessary condition implies that the solution must satisfy a fractional boundary value differential equation.

Let x have a continuous left Riemann–Liouville derivative of order α and J be a functional of the form

$$J[x] = \int_a^b L(t, x(t), {}_aD_t^\alpha x(t))dt \tag{9.1}$$

subject to the boundary conditions $x(a) = x_a$ and $x(b) = x_b$. Recall that

$$\begin{cases} \frac{\partial L}{\partial x} + {}_tD_b^\alpha \frac{\partial L}{\partial_a D_t^\alpha x} = 0 \\ x(a) = x_a, \quad x(b) = x_b \end{cases} \tag{9.2}$$

is the fractional Euler–Lagrange equation and is a necessary optimality condition.

Many variants of (9.2) can be found in the literature. Different types of fractional terms have been embedded in the Lagrangian and appropriate versions of Euler–Lagrange equations have been derived using proper integration by parts formulas. See (Agrawal, Muslih and Baleanu, 2011; Almeida, Pooseh and Torres, 2012; Atanacković, Konjik and Pilipović, 2008; Malinowska and Torres, 2010e; Odzijewicz, Malinowska and Torres, 2012a) for details.

For fractional optimal control problems, a so-called Hamiltonian system is constructed using Lagrange multipliers. For example, (cf. (Baleanu, Defterli and Agrawal, 2009)). Assume that we are required to minimize a functional of the form

$$J[x, u] = \int_a^b L(t, x(t), u(t)) dt$$

such that $x(a) = x_a$, $x(b) = x_b$ and $_aD_t^\alpha x(t) = f(t, x(t), u(t))$. Similar to the classical methods, one can introduce a Hamiltonian

$$H = L(t, x(t), u(t)) + \lambda(t) f(t, x(t), u(t)),$$

where $\lambda(t)$ is considered as a Lagrange multiplier. In this case we define the augmented functional as

$$J[x, u] = \int_a^b [H(t, x(t), u(t), \lambda(t)) - \lambda(t) {}_aD_t^\alpha x(t)] dt.$$

Optimizing the latter functional results into the following necessary optimality conditions:

$$\begin{cases} {}_aD_t^\alpha x(t) = \dfrac{\partial H}{\partial \lambda} \\ {}_tD_b^\alpha \lambda(t) = \dfrac{\partial H}{\partial x} \end{cases}, \qquad \dfrac{\partial H}{\partial u} = 0. \tag{9.3}$$

Together with the prescribed boundary conditions, this makes a two point fractional boundary value problem.

These arguments reveal that, like in the classical case, fractional variational problems end with fractional boundary value problems. To reach an optimal solution, one needs to deal with a fractional differential equation or a system of fractional differential equations. There are a few attempts in the literature to present analytic solutions to fractional variational problems. Simple problems have been treated in (Almeida and Torres, 2010); some other examples are presented in (Atanacković *et al.*, 2010).

Many solution methods, theoretical and numerical, furnish the classical theory of differential equations; nevertheless, solving a fractional differential equation is a rather tough task (Diethelm, 2010). To benefit from those methods, especially all solvers that are available to solve an integer-order differential equation numerically, we can either approximate a fractional variational problem by an equivalent integer-order one or approximate the necessary optimality conditions (9.2) and (9.3). The rest of this section discusses two types of approximations that are used to transform a fractional problem to one in which only integer-order derivatives are present, i.e., we approximate the original problem by substituting a fractional term by its corresponding expansion formulas. This is mainly done by case studies on certain examples. The examples are chosen so that either they have a trivial solution or it is possible to get an analytic solution using the fractional Euler–Lagrange equations (Pooseh, Almeida and Torres, 2013a).

By substituting the approximations (6.4) or (6.10) for the fractional derivative in (9.1), the problem is transformed to

$$J[x] \approx \int_a^b L\left(t, x(t), \sum_{k=0}^N \frac{(-1)^{k-1}\alpha x^{(k)}(t)}{k!(k-\alpha)\Gamma(1-\alpha)}(t-a)^{k-\alpha}\right) dt$$

$$= \int_a^b L'\left(t, x(t), \dot{x}(t), \dots, x^{(N)}(t)\right) dt$$

or

$$J[x] \approx \int_a^b L\left(t, x(t), \frac{Ax(t)}{(t-a)^\alpha} + \frac{B\dot{x}(t)}{(t-a)^{\alpha-1}} - \sum_{p=2}^N \frac{C(\alpha,p)V_p(t)}{(t-a)^{p+\alpha-1}}\right) dt$$

$$= \int_a^b L'\left(t, x(t), \dot{x}(t), V_2(t), \dots, V_N(t)\right) dt$$

with

$$\begin{cases} \dot{V}_p(t) = (1-p)(t-a)^{p-2}x(t) \\ V_p(a) = 0, \qquad p = 2, 3, \dots \end{cases}$$

The former problem is a classical variational problem containing higher order derivatives. The latter is a multistate problem, subject to an ordinary differential equation constraint. Together with the boundary conditions, both above problems belong to classes of well-studied variational problems.

To accomplish a detailed study, as test problems, we consider here Example 8.2,

$$\begin{cases} J[x] = \int_0^1 \left({}_0D_t^{0.5}x(t) - \dot{x}^2(t)\right) dt \to \min \\ x(0) = 0, \ x(1) = 1, \end{cases} \tag{9.4}$$

and the following example.

Example 9.1. Given $\alpha \in (0,1)$, consider the functional

$$J[x] = \int_0^1 \left({}_aD_t^\alpha x(t) - 1\right)^2 dt \tag{9.5}$$

to be minimized subject to the boundary conditions $x(0) = 0$ and $x(1) = \frac{1}{\Gamma(\alpha+1)}$. Since the integrand in (9.5) is non-negative, the functional attains its minimum when ${}_aD_t^\alpha x(t) = 1$, i.e., for $x(t) = \frac{t^\alpha}{\Gamma(\alpha+1)}$.

We illustrate the use of the two different expansions separately.

9.1   Expansion to integer orders

Using approximation (6.4) for the fractional derivative in (9.4), we get the approximated problem

$$\tilde{J}[x] = \int_0^1 \left[\sum_{n=0}^N C(n,\alpha) t^{n-\alpha} x^{(n)}(t) - \dot{x}^2(t) \right] dt \longrightarrow \min \qquad (9.6)$$

$$x(0) = 0, \quad x(1) = 1,$$

which is a classical higher-order problem of the calculus of variations that depends on derivatives up to order N. The corresponding necessary optimality condition is a well-known result.

Theorem 9.1 (cf., e.g., (Lebedev and Cloud, 2003)).
Suppose that $x \in C^{2N}[a,b]$ minimizes

$$\int_a^b L(t, x(t), x^{(1)}(t), x^{(2)}(t), \dots, x^{(N)}(t)) dt$$

with given boundary conditions

$$x(a) = a_0, \quad x(b) = b_0,$$
$$x^{(1)}(a) = a_1, \quad x^{(1)}(b) = b_1,$$
$$\vdots$$
$$x^{(N-1)}(a) = a_{N-1}, \quad x^{(N-1)}(b) = b_{N-1}.$$

Then x satisfies the Euler–Lagrange equation

$$\frac{\partial L}{\partial x} - \frac{d}{dt}\left(\frac{\partial L}{\partial x^{(1)}}\right) + \frac{d^2}{dt^2}\left(\frac{\partial L}{\partial x^{(2)}}\right) - \dots + (-1)^N \frac{d^N}{dt^N}\left(\frac{\partial L}{\partial x^{(N)}}\right) = 0. \quad (9.7)$$

In general (9.7) is an ODE (ordinary differential equation) of order $2N$, depending on the order N of the approximation we choose, and the method leaves $2N - 2$ parameters unknown. In our example, however, the Lagrangian in (9.6) is linear with respect to all derivatives of order higher than two. The resulting Euler–Lagrange equation is the second-order ODE

$$\sum_{n=0}^N (-1)^n C(n,\alpha) \frac{d^n}{dt^n}(t^{n-\alpha}) - \frac{d}{dt}[-2\dot{x}(t)] = 0,$$

that has the solution

$$x(t) = M_1(\alpha, N) t^{2-\alpha} + M_2(\alpha, N) t,$$

where

$$M_1(\alpha, N) = -\frac{1}{2\Gamma(3-\alpha)}\left[\sum_{n=0}^{N}(-1)^n\Gamma(n+1-\alpha)C(n,\alpha)\right],$$

$$M_2(\alpha, N) = \left[1 + \frac{1}{2\Gamma(3-\alpha)}\sum_{n=0}^{N}(-1)^n\Gamma(n+1-\alpha)C(n,\alpha)\right].$$

Figure 9.1 shows the analytic solution together with several approximations. It reveals that by increasing N, approximate solutions do not converge to the analytic one. The reason is the fact that the solution (8.13) to Example 8.2 is not an analytic function. We conclude that (6.4) may not be a good choice to approximate fractional variational problems. In contrast, as we shall see, the approximation (6.10) leads to good results.

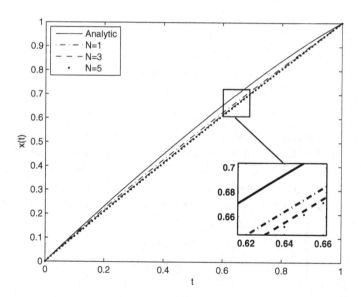

Fig. 9.1 Analytic *versus* approximate solutions to Example 8.2 using approximation (6.4) with $\alpha = 0.5$.

To solve Example 8.2 using (6.4) as an approximation for the fractional derivative, the problem becomes

$$\tilde{J}[x] = \int_0^1 \left(\sum_{n=0}^N C(n,\alpha) t^{n-\alpha} x^{(n)}(t) - 1 \right)^2 dt \longrightarrow \min$$

$$x(0) = 0, \quad x(1) = \frac{1}{\Gamma(\alpha+1)}.$$

The Euler–Lagrange equation (9.7) gives a $2N$ order ODE. For $N \geq 2$ this approach is inappropriate since the two given boundary conditions $x(0) = 0$ and $x(1) = \frac{1}{\Gamma(\alpha+1)}$ are not enough to determine the $2N$ constants of integration.

9.2   Expansion through the moments of a function

If we use (6.10) to approximate the optimization problem (9.4), with $A = A(\alpha, N)$, $B = B(\alpha, N)$ and $C_p = C(\alpha, p)$, we have

$$\tilde{J}[x] = \int_0^1 \left[At^{-\alpha} x(t) + Bt^{1-\alpha} \dot{x}(t) \right.$$

$$\left. - \sum_{p=2}^N C_p t^{1-p-\alpha} V_p(t) - \dot{x}^2(t) \right] dt \longrightarrow \min \tag{9.8}$$

$$\dot{V}_p(t) = (1-p)t^{p-2} x(t), \quad p = 2, 3, \ldots, N,$$

$$V_p(0) = 0, \quad p = 2, 3, \ldots, N,$$

$$x(0) = 0, \quad x(1) = 1.$$

Problem (9.8) is constrained with a set of ordinary differential equations and is natural to look to it as an optimal control problem (Pontryagin *et al.*, 1962). For that, we introduce the control variable $u(t) = \dot{x}(t)$. Then, using the Lagrange multipliers $\lambda_1, \lambda_2, \ldots, \lambda_N$, and the Hamiltonian system, one can reduce (9.8) to the study of the two point boundary value problem

$$\begin{cases} \dot{x}(t) = \frac{1}{2} Bt^{1-\alpha} - \frac{1}{2}\lambda_1(t), \\ \dot{V}_p(t) = (1-p)t^{p-2} x(t), \quad p = 2, 3, \ldots, N, \\ \dot{\lambda}_1(t) = At^{-\alpha} - \sum_{p=2}^N (1-p)t^{p-2}\lambda_p(t), \\ \dot{\lambda}_p(t) = -C_p t^{(1-p-\alpha)}, \quad p = 2, 3, \ldots, N, \end{cases} \tag{9.9}$$

with boundary conditions

$$\begin{cases} x(0) = 0, \\ V_p(0) = 0, \quad p = 2, 3, \ldots, N, \end{cases} \qquad \begin{cases} x(1) = 1, \\ \lambda_p(1) = 0, \quad p = 2, 3, \ldots, N, \end{cases}$$

where $x(0) = 0$ and $x(1) = 1$ are given. We have $V_p(0) = 0$, $p = 2, 3, \ldots, N$, due to (6.9) and $\lambda_p(1) = 0$, $p = 2, 3, \ldots, N$, because V_p is free at final time for $p = 2, 3, \ldots, N$ (Pontryagin *et al.*, 1962). In general, the Hamiltonian system is a nonlinear, hard to solve, two point boundary value problem that needs special numerical methods. In this case, however, (9.9) is a non-coupled system of ordinary differential equations and is easily solved to give

$$x(t) = M(\alpha, N)t^{2-\alpha} - \sum_{p=2}^{N} \frac{C(\alpha, p)}{2p(2 - p - \alpha)} t^p$$

$$+ \left[1 - M(\alpha, N) + \sum_{p=2}^{N} \frac{C(\alpha, p)}{2p(2 - p - \alpha)} \right] t,$$

where

$$M(\alpha, N) = \frac{1}{2(2 - \alpha)} \left[B(\alpha, N) - \frac{A(\alpha, N)}{1 - \alpha} - \sum_{p=2}^{N} \frac{C(\alpha, p)(1 - p)}{(1 - \alpha)(2 - p - \alpha)} \right].$$

Figure 9.2 shows the graph of x for different values of N.

Let us now approximate Example 9.1 using (6.10). The resulting minimization problem has the following form:

$$\tilde{J}[x] = \int_0^1 \left[At^{-\alpha}x(t) + Bt^{1-\alpha}\dot{x}(t) - \sum_{p=2}^{N} C_p t^{1-p-\alpha} V_p(t) - 1 \right]^2 dt \to \min$$

$$\dot{V}_p(t) = (1 - p)t^{p-2}x(t), \quad p = 2, 3, \ldots, N,$$

$$V_p(0) = 0, \quad p = 2, 3, \ldots, N,$$

$$x(0) = 0, \quad x(1) = \frac{1}{\Gamma(\alpha + 1)}.$$

(9.10)

Following the classical optimal control approach of Pontryagin (Pontryagin *et al.*, 1962), this time with

$$u(t) = At^{-\alpha}x(t) + Bt^{1-\alpha}\dot{x}(t) - \sum_{p=2}^{N} C_p t^{1-p-\alpha} V_p(t),$$

we conclude that the solution to (9.10) satisfies the system of differential equations

$$\begin{cases} \dot{x}(t) = -AB^{-1}t^{-1}x(t) + \sum_{p=2}^{N} B^{-1}C_p t^{-p}V_p(t) + \frac{1}{2}B^{-2}t^{2\alpha-2}\lambda_1(t) \\ \quad + B^{-1}t^{\alpha-1}, \\ \dot{V}_p(t) = (1 - p)t^{p-2}x(t), \quad p = 2, 3, \ldots, N, \\ \dot{\lambda}_1(t) = AB^{-1}t^{-1}\lambda_1 - \sum_{p=2}^{N}(1 - p)t^{p-2}\lambda_p(t), \\ \dot{\lambda}_p(t) = -B^{-1}C_p t^{-p}\lambda_1, \quad p = 2, 3, \ldots, N, \end{cases}$$

(9.11)

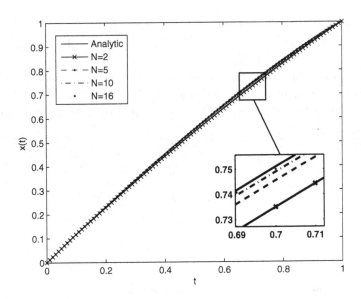

Fig. 9.2   Analytic *versus* approximate solutions to Example 8.2 using approximation (6.10) with $\alpha = 0.5$.

where $A = A(\alpha, N)$, $B = B(\alpha, N)$ and $C_p = C(\alpha, p)$ are defined according to Section 6.1.2, subject to the boundary conditions

$$
\begin{cases} x(0) = 0, \\ V_p(0) = 0, \quad p = 2, 3, \ldots, N, \end{cases} \qquad \begin{cases} x(1) = \dfrac{1}{\Gamma(\alpha + 1)}, \\ \lambda_p(1) = 0, \quad p = 2, 3, \ldots, N. \end{cases}
$$
$$(9.12)$$

The solution to system (9.11) and (9.12), with $N = 2$, is shown in Fig. 9.3.

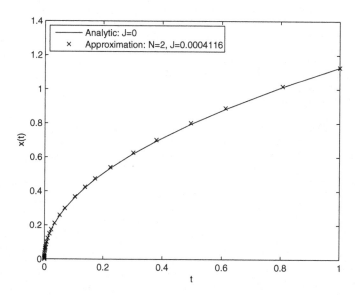

Fig. 9.3 Analytic *versus* approximate solution to Example 9.1 using approximation (6.10) with $\alpha = 0.5$.

Chapter 10

Fractional optimal control with free end-points

This chapter is devoted to fractional-order optimal control problems in which the dynamic control system involves integer and fractional-order derivatives and the terminal time is free. Necessary conditions for a state/control/terminal-time triplet to be optimal are obtained. Situations with constraints present at the end time are also considered. Under appropriate assumptions, it is shown that the obtained necessary optimality conditions become sufficient. Numerical methods to solve the problems are presented, and some computational simulations are discussed in detail (Pooseh, Almeida and Torres, 2014).

10.1 Necessary optimality conditions

Let $\alpha \in (0,1)$, $a \in \mathbb{R}$, L and f be two differentiable functions with domain $[a, +\infty) \times \mathbb{R}^2$, and $\phi : [a, +\infty) \times \mathbb{R} \to \mathbb{R}$ be a differentiable function. The fundamental problem is stated in the following way:

$$J[x, u, T] = \int_a^T L(t, x(t), u(t)) \, dt + \phi(T, x(T)) \longrightarrow \min \qquad (10.1)$$

subject to the control system

$$M\dot{x}(t) + N \, {}_a^C D_t^\alpha x(t) = f(t, x(t), u(t)), \qquad (10.2)$$

and the initial boundary condition

$$x(a) = x_a, \qquad (10.3)$$

with $(M, N) \neq (0,0)$ and x_a a fixed real number. Our goal is to generalize previous works on fractional optimal control problems by considering the end time, T, free and the dynamic control system (10.2) involving integer and fractional-order derivatives. For convenience, we consider the one-dimensional case. However, using similar techniques, the results can be easily extended to problems with multiple states and multiple controls. Later we consider the cases T and/or $x(T)$ fixed. Here, T is a variable number with $a < T < \infty$. Thus, we are interested not only in the optimal trajectory x and optimal control function u, but also in the corresponding time T for which the functional J attains its minimum value. We assume that the state variable x is differentiable and that the control u is piecewise continuous. When $N = 0$ we obtain a classical optimal control problem; the case $M = 0$ with fixed T has already been studied for different types of fractional-order derivatives (see, e.g., (Agrawal, 2004, 2008a; Agrawal, Defterli and Baleanu, 2010; Frederico and Torres, 2008b,a; Tricaud and Chen, 2010a,b)). In (Jelicic and Petrovacki, 2009) a special type of the proposed problem is also studied for fixed T.

Remark 10.1. In this chapter the terminal time T is a free decision variable and, *a priori*, no constraints are imposed. For future research, one may wish to consider a class of fractional optimal control problems in which the terminal time is governed by a stopping condition. Such problems were recently investigated, within the classical (integer-order) framework, in (Lin *et al.*, 2011, 2012).

10.1.1 *Fractional necessary conditions*

To deduce necessary optimality conditions that an optimal triplet (x, u, T) must satisfy, we use a Lagrange multiplier to adjoin the dynamic constraint (10.2) to the performance functional (10.1). To start, we define the Hamiltonian function H by

$$H(t, x, u, \lambda) = L(t, x, u) + \lambda f(t, x, u), \qquad (10.4)$$

where λ is a Lagrange multiplier, so that we can rewrite the initial problem as minimizing

$$\mathcal{J}[x, u, T, \lambda] = \int_a^T \left[H(t, x(t), u(t), \lambda(t)) - \lambda(t)[M\dot{x}(t) + N \, {}_a^C D_t^\alpha x(t)] \right] dt$$
$$+ \phi(T, x(T)).$$

Next, we consider variations of the form

$$x + \delta x, \quad u + \delta u, \quad T + \delta T, \quad \lambda + \delta \lambda,$$

with $\delta x(a) = 0$ by the imposed boundary condition (10.3). Using the well-known fact that the first variation of $\mathcal{J}$ must vanish when evaluated along a minimizer, we get

$$0 = \int_a^T \left(\frac{\partial H}{\partial x} \delta x + \frac{\partial H}{\partial u} \delta u + \frac{\partial H}{\partial \lambda} \delta \lambda - \delta \lambda \left(M \dot{x}(t) + N \, {}_a^C D_t^\alpha x(t) \right) \right.$$

$$\left. - \lambda(t) \left(M \dot{\delta x}(t) + N \, {}_a^C D_t^\alpha \delta x(t) \right) \right) dt + \delta T \left[H(t, x, u, \lambda) \right.$$

$$\left. - \lambda(t) \left(M \dot{x}(t) + N \, {}_a^C D_t^\alpha x(t) \right) \right]_{t=T} + \frac{\partial \phi}{\partial t}(T, x(T)) \delta T$$

$$+ \frac{\partial \phi}{\partial x}(T, x(T)) \left(\dot{x}(T) \delta T + \delta x(T) \right)$$

with the partial derivatives of H evaluated at $(t, x(t), u(t), \lambda(t))$. Integration by parts gives the relations

$$\int_a^T \lambda(t) \dot{\delta x}(t) \, dt = - \int_a^T \delta x(t) \dot{\lambda}(t) \, dt + \delta x(T) \lambda(T)$$

and

$$\int_a^T \lambda(t) \, {}_a^C D_t^\alpha \delta x(t) \, dt = \int_a^T \delta x(t) \, {}_t D_T^\alpha \lambda(t) \, dt + \delta x(T) [{}_t I_T^{1-\alpha} \lambda(t)]_{t=T}.$$

Thus, we deduce the following formula:

$$\int_a^T \left[\delta x \left(\frac{\partial H}{\partial x} + M \dot{\lambda} - N \, {}_t D_T^\alpha \lambda \right) + \delta u \frac{\partial H}{\partial u} \right.$$

$$\left. + \delta \lambda \left(\frac{\partial H}{\partial \lambda} - M \dot{x} - N \, {}_a^C D_t^\alpha x \right) \right] dt - \delta x(T) \left[M \lambda + N \, {}_t I_T^{1-\alpha} \lambda - \frac{\partial \phi}{\partial x}(t, x) \right]_{t=T}$$

$$+ \delta T \left[H(t, x, u, \lambda) - \lambda [M \dot{x} + N \, {}_a^C D_t^\alpha x] + \frac{\partial \phi}{\partial t}(t, x) + \frac{\partial \phi}{\partial x}(t, x) \dot{x} \right]_{t=T} = 0.$$

Now, define the new variable

$$\delta x_T = [x + \delta x](T + \delta T) - x(T).$$

Because $\dot{\delta x}(T)$ is arbitrary, in particular one can consider variation functions for which $\dot{\delta x}(T) = 0$. By Taylor's theorem,

$$[x + \delta x](T + \delta T) - [x + \delta x](T) = \dot{x}(T) \delta T + O(\delta T^2),$$

where $\lim\limits_{\zeta \to 0} \dfrac{O(\zeta)}{\zeta}$ is finite, and so $\delta x(T) = \delta x_T - \dot{x}(T)\delta T + O(\delta T^2)$. In conclusion, we arrive at the expression

$$\delta T \left[H(t, x, u, \lambda) - N\lambda(t)\,_a^C D_t^\alpha x(t) + N\dot{x}(t)\,_t I_T^{1-\alpha}\lambda(t) + \frac{\partial \phi}{\partial t}(t, x(t)) \right]_{t=T}$$

$$+ \int_a^T \left[\delta x \left(\frac{\partial H}{\partial x} + M\dot{\lambda}(t) - N\,_t D_T^\alpha \lambda(t) \right) \right.$$

$$+ \delta\lambda \left(\frac{\partial H}{\partial \lambda} - M\dot{x}(t) - N\,_a^C D_t^\alpha x(t) \right) + \delta u \frac{\partial H}{\partial u} \right] dt$$

$$- \delta x_T \left[M\lambda(t) + N\,_t I_T^{1-\alpha}\lambda(t) - \frac{\partial \phi}{\partial x}(t, x(t)) \right]_{t=T} + O(\delta T^2) = 0.$$

Since the variation functions were chosen arbitrarily, the following theorem is proven. Compare with (Kamocki, 2014).

Theorem 10.1. *If (x, u, T) is a minimizer of (10.1) under the dynamic constraint (10.2) and the boundary condition (10.3), then there exists a function λ for which the triplet (x, u, λ) satisfies:*

- *the Hamiltonian system*

$$\begin{cases} M\dot{\lambda}(t) - N\,_t D_T^\alpha \lambda(t) = -\dfrac{\partial H}{\partial x}(t, x(t), u(t), \lambda(t)) \\[2mm] M\dot{x}(t) + N\,_a^C D_t^\alpha x(t) = \dfrac{\partial H}{\partial \lambda}(t, x(t), u(t), \lambda(t)) \end{cases} \tag{10.5}$$

  for all $t \in [a, T]$;
- *the stationary condition*

$$\frac{\partial H}{\partial u}(t, x(t), u(t), \lambda(t)) = 0 \tag{10.6}$$

  for all $t \in [a, T]$;
- *and the transversality conditions*

$$\left[H(t, x(t), u(t), \lambda(t)) - N\lambda(t)\,_a^C D_t^\alpha x(t) + N\dot{x}(t)\,_t I_T^{1-\alpha}\lambda(t) \right.$$

$$\left. + \frac{\partial \phi}{\partial t}(t, x(t)) \right]_{t=T} = 0, \tag{10.7}$$

$$\left[M\lambda(t) + N\,_t I_T^{1-\alpha}\lambda(t) - \frac{\partial \phi}{\partial x}(t, x(t)) \right]_{t=T} = 0;$$

where the Hamiltonian H is defined by (10.4).

Remark 10.2. In standard optimal control, a free terminal time problem can be converted into a fixed final time problem by using the well-known transformation $s = t/T$ (see Example 10.2). This transformation does not work in the fractional setting. Indeed, in standard optimal control, translating the problem from time t to a new time variable s is straightforward: the chain rule gives $\frac{dx}{ds} = \frac{dx}{dt}\frac{dt}{ds}$. For Caputo or Riemann–Liouville fractional derivatives, the chain rule has no practical use and such conversion is not possible.

Some interesting special cases are obtained when restrictions are imposed on the end time T or on $x(T)$.

Corollary 10.1. *Let (x, u) be a minimizer of* (10.1) *under the dynamic constraint* (10.2) *and the boundary condition* (10.3).

(1) If T is fixed and $x(T)$ is free, then Theorem 10.1 holds with the transversality conditions (10.7) *replaced by*

$$\left[M\lambda(t) + N\,{}_tI_T^{1-\alpha}\lambda(t) - \frac{\partial\phi}{\partial x}(t, x(t)) \right]_{t=T} = 0.$$

(2) If $x(T)$ is fixed and T is free, then Theorem 10.1 holds with the transversality conditions (10.7) *replaced by*

$$\left[H(t, x(t), u(t), \lambda(t)) - N\lambda(t)_a^C D_t^\alpha x(t) + N\dot{x}(t)_t I_T^{1-\alpha}\lambda(t) \right.$$
$$\left. + \frac{\partial\phi}{\partial t}(t, x(t)) \right]_{t=T} = 0.$$

(3) If T and $x(T)$ are both fixed, then Theorem 10.1 holds with no transversality conditions.

(4) If the terminal point $x(T)$ belongs to a fixed curve, i.e., $x(T) = \gamma(T)$ for some differentiable curve γ, then Theorem 10.1 holds with the transversality conditions (10.7) *replaced by*

$$\left[H(t, x(t), u(t), \lambda(t)) - N\lambda(t)_a^C D_t^\alpha x(t) + N\dot{x}(t)_t I_T^{1-\alpha}\lambda(t) + \frac{\partial\phi}{\partial t}(t, x(t)) \right.$$
$$\left. - \dot{\gamma}(t)\left(M\lambda(t) + N\,{}_tI_T^{1-\alpha}\lambda(t) - \frac{\partial\phi}{\partial x}(t, x(t)) \right) \right]_{t=T} = 0.$$

(5) *If T is fixed and $x(T) \geq K$ for some fixed $K \in \mathbb{R}$, then Theorem 10.1 holds with the transversality conditions (10.7) replaced by*

$$\left[M\lambda(t) + N\,_tI_T^{1-\alpha}\lambda(t) - \frac{\partial \phi}{\partial x}(t, x(t)) \right]_{t=T} \leq 0,$$

$$(x(T) - K)\left[M\lambda(t) + N\,_tI_T^{1-\alpha}\lambda(t) - \frac{\partial \phi}{\partial x}(t, x(t)) \right]_{t=T} = 0.$$

(6) *If $x(T)$ is fixed and $T \leq K$ for some fixed $K \in \mathbb{R}$, then Theorem 10.1 holds with the transversality conditions (10.7) replaced by*

$$\left[H(t, x(t), u(t), \lambda(t)) - N\lambda(t)_a^C D_t^\alpha x(t) + N\dot{x}(t)\,_tI_T^{1-\alpha}\lambda(t) \right.$$

$$\left. + \frac{\partial \phi}{\partial t}(t, x(t)) \right]_{t=T} \geq 0$$

and

$$\left[H(t, x(t), u(t), \lambda(t)) - N\lambda(t)_a^C D_t^\alpha x(t) + N\dot{x}(t)\,_tI_T^{1-\alpha}\lambda(t) \right.$$

$$\left. + \frac{\partial \phi}{\partial t}(t, x(t)) \right]_{t=T} (T - K) = 0.$$

Proof.   The first three conditions are obvious. The fourth follows from

$$\delta x_T = \gamma(T + \delta T) - \gamma(T) = \dot{\gamma}(T)\delta T + O(\delta T^2).$$

To prove *(5)*, observe that we have two possible cases. If $x(T) > K$, then δx_T may take negative and positive values, and so we get

$$\left[M\lambda(t) + N\,_tI_T^{1-\alpha}\lambda(t) - \frac{\partial \phi}{\partial x}(t, x(t)) \right]_{t=T} = 0.$$

On the other hand, if $x(T) = K$, then $\delta x_T \geq 0$ and so by the Karush–Kuhn–Tucker theorem

$$\left[M\lambda(t) + N\,_tI_T^{1-\alpha}\lambda(t) - \frac{\partial \phi}{\partial x}(t, x(t)) \right]_{t=T} \leq 0.$$

The proof of the last condition is similar.   □

Case 1 of Corollary 10.1 was proven in (Frederico and Torres, 2008b) for $(M, N) = (0, 1)$ and $\phi \equiv 0$. Moreover, if $\alpha = 1$, then we obtain the classical necessary optimality conditions for the standard optimal control problem (see, e.g., (Chiang, 1992)):

- the Hamiltonian system

$$\begin{cases} \dot{x}(t) = \dfrac{\partial H}{\partial \lambda}(t, x(t), u(t), \lambda(t)), \\[2mm] \dot{\lambda}(t) = -\dfrac{\partial H}{\partial x}(t, x(t), u(t), \lambda(t)), \end{cases}$$

- the stationary condition

$$\frac{\partial H}{\partial u}(t, x(t), u(t), \lambda(t)) = 0,$$

- the transversality condition $\lambda(T) = 0$.

10.1.2 Approximated integer-order necessary optimality conditions

Using approximation (6.10), and the relation between Caputo and Riemann–Liouville derivatives, up to order K, we can transform the original problem (10.1) to (10.3) into the following classical problem:

$$\tilde{J}[x, u, T] = \int_a^T L(t, x(t), u(t)) \, dt + \phi(T, x(T)) \longrightarrow \min$$

subject to

$$\begin{cases} \dot{x}(t) = \dfrac{1}{M + NB(t-a)^{1-\alpha}} \left(f(t, x(t), u(t)) - NA(t-a)^{-\alpha}x(t) \right. \\[3mm] \qquad\qquad \left. + \sum_{p=2}^{K} NC_p(t-a)^{1-p-\alpha}V_p(t) - \dfrac{x(a)(t-a)^{-\alpha}}{\Gamma(1-\alpha)} \right) \\[3mm] \dot{V}_p(t) = (1-p)(t-a)^{p-2}x(t), \quad p = 2, \ldots, K, \end{cases}$$

and

$$\begin{cases} x(a) = x_a, \\ V_p(a) = 0, \quad p = 2, \ldots, K, \end{cases} \tag{10.8}$$

where $A = A(\alpha, K)$, $B = B(\alpha, K)$ and $C_p = C(\alpha, p)$ are the coefficients in the approximation (6.10). Now that we are dealing with an integer-order problem, we can follow a classical procedure (see, e.g., (Kirk, 1970)), by defining the Hamiltonian H by

$$H = L(t, x, u) + \frac{\lambda_1}{M + NB(t-a)^{1-\alpha}} \left(f(t, x, u) - NA(t-a)^{-\alpha}x \right.$$

$$+ \sum_{p=2}^{K} NC_p(t-a)^{1-p-\alpha}V_p - \frac{x(a)(t-a)^{-\alpha}}{\Gamma(1-\alpha)} \right) + \sum_{p=2}^{K} \lambda_p(1-p)(t-a)^{p-2}x.$$

Let $\boldsymbol{\lambda} = (\lambda_1, \lambda_2, \ldots, \lambda_K)$ and $\mathbf{x} = (x, V_2, \ldots, V_K)$. The necessary optimality conditions

$$\frac{\partial H}{\partial u} = 0, \qquad \begin{cases} \dot{\mathbf{x}} = \dfrac{\partial H}{\partial \boldsymbol{\lambda}}, \\[2mm] \dot{\boldsymbol{\lambda}} = -\dfrac{\partial H}{\partial \mathbf{x}}, \end{cases}$$

result in a two point boundary value problem. Assume that $(T^*, \mathbf{x}^*, \mathbf{u}^*)$ is the optimal triplet. In addition to the boundary conditions (10.8), the transversality conditions imply

$$\left[H(T^*, \mathbf{x}^*(T), \mathbf{u}^*(T), \boldsymbol{\lambda}^*(T)) + \frac{\partial \phi}{\partial t}(T^*, \mathbf{x}^*(T)) \right] \delta T$$
$$+ \left[\boldsymbol{\lambda}(T) - \frac{\partial \phi}{\partial \mathbf{x}}(T^*, \mathbf{x}^*(T)) \right]^{tr} \delta \mathbf{x}_T = 0,$$

where tr denotes the transpose. Because V_p, $p = 2, \ldots, K$, are auxiliary variables whose values $V_p(T)$, at the final time T, are free, we have

$$\lambda_p(T) = \left. \frac{\partial \phi}{\partial V_p} \right|_{t=T} = 0, \quad p = 2, \ldots, K.$$

The value of $\lambda_1(T)$ is determined from the value of $x(T)$. If $x(T)$ is free, then $\lambda_1(T) = \frac{\partial \phi}{\partial x}|_{t=T}$. Whenever the final time is free, a transversality condition of the form

$$\left[H\left(t, \mathbf{x}(t), \mathbf{u}(t), \boldsymbol{\lambda}(t)\right) - \frac{\partial \phi}{\partial t}\left(t, \mathbf{x}(t)\right) \right]_{t=T} = 0$$

completes the required set of boundary conditions.

10.2   A generalization

The aim is now to consider a generalization of the optimal control problem (10.1) to (10.3) studied in Section 10.1. Observe that the initial point $t = a$ is in fact the initial point for two different operators: for the integral in (10.1) and, secondly, for the left Caputo fractional derivative given by the dynamic constraint (10.2). We now consider the case where the lower bound of the integral of J is greater than the lower bound of the fractional derivative. The problem is stated as follows:

$$J[x, u, T] = \int_A^T L(t, x(t), u(t))\, dt + \phi(T, x(T)) \longrightarrow \min \qquad (10.9)$$

under the constraints

$$M\dot{x}(t) + N\, {}_a^C D_t^\alpha x(t) = f(t, x(t), u(t)) \quad \text{and} \quad x(A) = x_A, \qquad (10.10)$$

where $(M, N) \neq (0, 0)$, x_A is a fixed real, and $a < A$.

Remark 10.3. We have chosen to consider the initial condition on the initial time A of the cost integral, but the case of initial condition $x(a)$ instead of $x(A)$ can be studied using similar arguments. Our choice seems the most natural: the interval of interest is $[A, T]$ but the fractional derivative is a non-local operator and has "memory" that goes to the past of the interval $[A, T]$ under consideration.

Remark 10.4. In the theory of fractional differential equations, the initial condition is given at $t = a$. To the best of our knowledge there is no general theory about uniqueness of solutions for problems like (10.10), where the fractional derivative involves $x(t)$ for $a < t < A$ and the initial condition is given at $t = A$. Uniqueness of solution is, however, possible. Consider, for example, ${}_0^C D_t^\alpha x(t) = t^2$. Applying the fractional integral to both sides of equality we get $x(t) = x(0) + 2t^{2+\alpha}/\Gamma(3 + \alpha)$ so, knowing a value for $x(t)$, not necessarily at $t = 0$, one can determine $x(0)$ and by doing so $x(t)$. A different approach than the one considered here is to provide an initialization function for $t \in [a, A]$. This initial memory approach was studied for fractional continuous-time linear control systems in (Mozyrska and Torres, 2010) and (Mozyrska and Torres, 2011), respectively for Caputo and Riemann–Liouville derivatives.

The method to obtain the required necessary optimality conditions follows the same procedure as the one discussed before. The first variation gives

$$0 = \int_A^T \left[\frac{\partial H}{\partial x} \delta x + \frac{\partial H}{\partial u} \delta u + \frac{\partial H}{\partial \lambda} \delta \lambda - \delta \lambda \left(M\dot{x}(t) + N\, {}_a^C D_t^\alpha x(t) \right) \right.$$

$$\left. - \lambda(t) \left(M\dot{\delta x}(t) + N\, {}_a^C D_t^\alpha \delta x(t) \right) \right] dt + \frac{\partial \phi}{\partial x}(T, x(T)) \left(\dot{x}(T)\delta T + \delta x(T) \right)$$

$$+ \frac{\partial \phi}{\partial t}(T, x(T))\delta T + \delta T \left[H(t, x, u, \lambda) - \lambda(t) \left(M\dot{x}(t) + N\, {}_a^C D_t^\alpha x(t) \right) \right]_{t=T},$$

where the Hamiltonian H is as in (10.4). Now, if we integrate by parts, we get

$$\int_A^T \lambda(t)\dot{\delta x}(t)\, dt = -\int_A^T \delta x(t)\dot{\lambda}(t)\, dt + \delta x(T)\lambda(T),$$

and

$$\int_A^T \lambda(t)\,{}_a^C D_t^\alpha \delta x(t)\,dt = \int_a^T \lambda(t)\,{}_a^C D_t^\alpha \delta x(t)\,dt - \int_a^A \lambda(t)\,{}_a^C D_t^\alpha \delta x(t)\,dt$$

$$= \int_a^T \delta x(t)\,{}_t D_T^\alpha \lambda(t)\,dt + [\delta x(t)\,{}_t I_T^{1-\alpha} \lambda(t)]_{t=a}^{t=T} - \int_a^A \delta x(t)\,{}_t D_A^\alpha \lambda(t)\,dt$$

$$- [\delta x(t)\,{}_t I_A^{1-\alpha} \lambda(t)]_{t=a}^{t=A}$$

$$= \int_a^A \delta x(t)[{}_t D_T^\alpha \lambda(t) - {}_t D_A^\alpha \lambda(t)]\,dt + \int_A^T \delta x(t)\,{}_t D_T^\alpha \lambda(t)\,dt$$

$$+ \delta x(T)[{}_t I_T^{1-\alpha} \lambda(t)]_{t=T} - \delta x(a)[{}_a I_T^{1-\alpha} \lambda(a) - {}_a I_A^{1-\alpha} \lambda(a)].$$

Substituting these relations into the first variation of J, we conclude that

$$\int_A^T \left(\frac{\partial H}{\partial x} + M\dot\lambda - N\,{}_t D_T^\alpha \lambda \right) \delta x + \frac{\partial H}{\partial u} \delta u + \left(\frac{\partial H}{\partial \lambda} - M\dot x - N\,{}_a^C D_t^\alpha x \right) \delta\lambda\,dt$$

$$- N \int_a^A \delta x[{}_t D_T^\alpha \lambda - {}_t D_A^\alpha \lambda]\,dt - \delta x[M\lambda + N\,{}_t I_T^{1-\alpha}\lambda - \frac{\partial \phi}{\partial x}(t,x)]_{t=T}$$

$$+ \delta T[H(t,x,u,\lambda) - \lambda[M\dot x + N\,{}_a^C D_t^\alpha x] + \frac{\partial \phi}{\partial t}(t,x) + \frac{\partial \phi}{\partial x}(t,x)\dot x]_{t=T}$$

$$+ N\delta x(a)[{}_a I_T^{1-\alpha}\lambda(a) - {}_a I_A^{1-\alpha}\lambda(a)] = 0.$$

Repeating the calculations as before, we prove the following optimality conditions.

Theorem 10.2. *If the triplet (x, u, T) is an optimal solution to problem (10.9) and (10.10), then there exists a function λ for which the following conditions hold:*

- *the Hamiltonian system*

$$\begin{cases} M\dot\lambda(t) - N\,{}_t D_T^\alpha \lambda(t) = -\dfrac{\partial H}{\partial x}(t, x(t), u(t), \lambda(t)) \\[2mm] M\dot x(t) + N\,{}_a^C D_t^\alpha x(t) = \dfrac{\partial H}{\partial \lambda}(t, x(t), u(t), \lambda(t)) \end{cases}$$

for all $t \in [A, T]$, and ${}_t D_T^\alpha \lambda(t) - {}_t D_A^\alpha \lambda(t) = 0$ for all $t \in [a, A]$;
- *the stationary condition*

$$\frac{\partial H}{\partial u}(t, x(t), u(t), \lambda(t)) = 0$$

for all $t \in [A, T]$;

- *the transversality conditions*

$$\left[H(t, x(t), u(t), \lambda(t)) - N\lambda(t) {}_a^C D_t^\alpha x(t) + N\dot{x}(t)_t I_T^{1-\alpha} \lambda(t) \right.$$

$$\left. + \frac{\partial \phi}{\partial t}(t, x(t)) \right]_{t=T} = 0,$$

$$\left[M\lambda(t) + N {}_t I_T^{1-\alpha} \lambda(t) - \frac{\partial \phi}{\partial x}(t, x(t)) \right]_{t=T} = 0,$$

$$\left[{}_t I_T^{1-\alpha} \lambda(t) - {}_t I_A^{1-\alpha} \lambda(t) \right]_{t=a} = 0;$$

with the Hamiltonian H given by (10.4).

Remark 10.5. If the admissible functions take fixed values at both $t = a$ and $t = A$, then we only obtain the two transversality conditions evaluated at $t = T$.

10.3 Sufficient optimality conditions

In this section we show that, under some extra hypotheses, the obtained necessary optimality conditions are also sufficient. To begin, let us recall the notions of convexity and concavity for C^1 functions of several variables. We refer the reader to Section 2.4 of (Malinowska and Torres, 2012).

Definition 10.1. *Given $k \in \{1, \ldots, n\}$ and a function $\Psi : D \subseteq \mathbb{R}^n \to \mathbb{R}$ such that $\partial \Psi / \partial t_i$ exist and are continuous for all $i \in \{k, \ldots, n\}$, we say that Ψ is convex (concave) in $(t_k, \ldots, t_n)$ if*

$$\Psi(t_1 + \theta_1, \ldots, t_{k-1} + \theta_{k-1}, t_k + \theta_k, \ldots, t_n + \theta_n) - \Psi(t_1, \ldots, t_{k-1}, t_k, \ldots, t_n)$$

$$\geq (\leq) \frac{\partial \Psi}{\partial t_k}(t_1, \ldots, t_{k-1}, t_k, \ldots, t_n)\theta_k + \cdots + \frac{\partial \Psi}{\partial t_n}(t_1, \ldots, t_{k-1}, t_k, \ldots, t_n)\theta_n$$

for all $(t_1, \ldots, t_n), (t_1 + \theta_1, \ldots, t_n + \theta_n) \in D$.

Theorem 10.3. *Let $(\overline{x}, \overline{u}, \overline{\lambda})$ be a triplet satisfying conditions (10.5) to (10.7) of Theorem 10.1. Moreover, assume that*

(1) L and f are convex on x and u, and ϕ is convex in x;
(2) T is fixed;
(3) $\overline{\lambda}(t) \geq 0$ for all $t \in [a, T]$ or f is linear in x and u.

Then $(\overline{x}, \overline{u})$ is an optimal solution to problem (10.1) to (10.3).

Proof.　From (10.5) we deduce that

$$\frac{\partial L}{\partial x}(t, \overline{x}(t), \overline{u}(t)) = -M\dot{\overline{\lambda}}(t) + N \, {}_tD_T^\alpha\overline{\lambda}(t) - \overline{\lambda}(t)\frac{\partial f}{\partial x}(t, \overline{x}(t), \overline{u}(t)).$$

Using (10.6),

$$\frac{\partial L}{\partial u}(t, \overline{x}(t), \overline{u}(t)) = -\overline{\lambda}(t)\frac{\partial f}{\partial u}(t, \overline{x}(t), \overline{u}(t)),$$

and (10.7) gives $[M\overline{\lambda}(t) + N \, {}_tI_T^{1-\alpha}\overline{\lambda}(t) - \frac{\partial \phi}{\partial x}(t, \overline{x}(t))]_{t=T} = 0$. Let (x, u) be admissible, i.e., let (10.2) and (10.3) be satisfied for (x, u). In this case,

$$\triangle J = J[x, u] - J[\overline{x}, \overline{u}]$$

$$= \int_a^T [L(t, x(t), u(t)) - L(t, \overline{x}(t), \overline{u}(t))] \, dt + \phi(T, x(T)) - \phi(T, \overline{x}(T))$$

$$\geq \int_a^T \frac{\partial L}{\partial x}(t, \overline{x}(t), \overline{u}(t))(x(t) - \overline{x}(t)) + \frac{\partial L}{\partial u}(t, \overline{x}(t), \overline{u}(t))(u(t) - \overline{u}(t)) \, dt$$

$$+ \frac{\partial \phi}{\partial x}(T, \overline{x}(T))(x(T) - \overline{x}(T))$$

$$= \int_a^T \left[-M\dot{\overline{\lambda}}(t)(x(t) - \overline{x}(t)) + N \, {}_tD_T^\alpha\overline{\lambda}(t)(x(t) - \overline{x}(t)) \right.$$

$$- \overline{\lambda}(t)\frac{\partial f}{\partial x}(t, \overline{x}(t), \overline{u}(t))(x(t) - \overline{x}(t))$$

$$\left. - \overline{\lambda}(t)\frac{\partial f}{\partial u}(t, \overline{x}(t), \overline{u}(t))(u(t) - \overline{u}(t)) \right] dt$$

$$+ \frac{\partial \phi}{\partial x}(T, \overline{x}(T))(x(T) - \overline{x}(T)).$$

Integrating by parts, and noting that $x(a) = \overline{x}(a)$, we obtain

$$\triangle J \geq \int_a^T \overline{\lambda}(t) \left[M\left(\dot{x}(t) - \dot{\overline{x}}(t)\right) + N \left({}_a^CD_t^\alpha x(t) - {}_a^CD_t^\alpha\overline{x}(t)\right) \right.$$

$$- \frac{\partial f}{\partial x}(t, \overline{x}(t), \overline{u}(t))(x(t) - \overline{x}(t))$$

$$\left. - \frac{\partial f}{\partial u}(t, \overline{x}(t), \overline{u}(t))(u(t) - \overline{u}(t)) \right] dt$$

$$+ \left[\frac{\partial \phi}{\partial x}(t, \overline{x}(t)) - M\overline{\lambda}(t) - N \, {}_tI_T^{1-\alpha}\overline{\lambda}(t) \right]_{t=T} (x(T) - \overline{x}(T)),$$

and finally

$$\triangle J \geq \int_a^T \left[\overline{\lambda}(t) \left[f(t, x(t), u(t)) - f(t, \overline{x}(t), \overline{u}(t)) \right] \right.$$

$$- \overline{\lambda}(t) \frac{\partial f}{\partial x}(t, \overline{x}(t), \overline{u}(t)) (x(t) - \overline{x}(t))$$

$$\left. - \overline{\lambda}(t) \frac{\partial f}{\partial u}(t, \overline{x}(t), \overline{u}(t)) (u(t) - \overline{u}(t)) \right] dt$$

$$\geq \int_a^T \overline{\lambda}(t) \left[\frac{\partial f}{\partial x}(t, \overline{x}(t), \overline{u}(t)) (x(t) - \overline{x}(t)) + \frac{\partial f}{\partial u}(t, \overline{x}(t), \overline{u}(t)) (u(t) - \overline{u}(t)) \right.$$

$$\left. - \frac{\partial f}{\partial x}(t, \overline{x}(t), \overline{u}(t)) (x(t) - \overline{x}(t)) - \frac{\partial f}{\partial u}(t, \overline{x}(t), \overline{u}(t)) (u(t) - \overline{u}(t)) \right] dt = 0.$$
$\square$

Remark 10.6. If the functions in Theorem 10.3 are strictly convex, instead of convex, then the minimizer is unique.

10.4   Numerical treatment and examples

Here we apply the necessary conditions of Section 10.1 to solve some test problems. Solving an optimal control problem, analytically, is an optimistic goal and is impossible except for simple cases. Therefore, we apply numerical and computational methods to solve our problems. In each case we try to solve the problem either by applying fractional necessary conditions or by approximating the problem by a classical one and then solving the approximate problem.

10.4.1   Fixed final time

We first solve a simple problem with fixed final time. In this case the exact solution, i.e., the optimal control and the corresponding optimal trajectory, is known, and hence we can compare it with the approximations obtained by our numerical method.

Example 10.1. Consider the following optimal control problem:

$$J[x, u] = \int_0^1 \left(tu(t) - (\alpha + 2)x(t) \right)^2 dt \longrightarrow \min$$

subject to the control system

$$\dot{x}(t) + {}_0^C D_t^\alpha x(t) = u(t) + t^2,$$

and the boundary conditions

$$x(0) = 0, \quad x(1) = \frac{2}{\Gamma(3+\alpha)}.$$

The solution is given by

$$(\overline{x}(t), \overline{u}(t)) = \left(\frac{2t^{\alpha+2}}{\Gamma(\alpha+3)}, \frac{2t^{\alpha+1}}{\Gamma(\alpha+2)} \right),$$

because $J[x, u] \geq 0$ for all pairs (x, u) and $\overline{x}(0) = 0$, $\overline{x}(1) = \frac{2}{\Gamma(3+\alpha)}$, $\dot{\overline{x}}(t) = \overline{u}(t)$ and $_0^C D_t^\alpha \overline{x}(t) = t^2$ with $J[\overline{x}, \overline{u}] = 0$. It is trivial to check that $(\overline{x}, \overline{u})$ satisfies the fractional necessary optimality conditions given by Theorem 10.1/Corollary 10.1.

Let us apply the fractional necessary conditions to the above problem. The Hamiltonian is $H = (tu - (\alpha+2)x)^2 + \lambda u + \lambda t^2$. The stationary condition (10.6) implies that for $t \neq 0$

$$u(t) = \frac{\alpha+2}{t}x(t) - \frac{\lambda(t)}{2t^2},$$

and hence

$$H = -\frac{\lambda^2}{4t^2} + \frac{\alpha+2}{t}x\lambda + t^2\lambda, \quad t \neq 0. \tag{10.11}$$

Finally, (10.5) gives

$$\begin{cases} \dot{x}(t) + {}_0^C D_t^\alpha x(t) = -\dfrac{\lambda}{2t^2} + \dfrac{\alpha+2}{t}x(t) + t^2 \\ -\dot{\lambda}(t) + {}_t D_1^\alpha \lambda(t) = \dfrac{\alpha+2}{t}\lambda(t), \end{cases} \qquad \begin{cases} x(0) = 0 \\ x(1) = \dfrac{2}{\Gamma(3+\alpha)}. \end{cases}$$

At this point, we encounter a fractional boundary value problem that needs to be solved in order to reach the optimal solution. A handful of methods can be found in the literature to solve this problem. Nevertheless, we use approximations (6.10) and (6.14), up to order N, that have been introduced in (Atanacković and Stanković, 2008) and used in (Jelicic and Petrovacki, 2009; Pooseh, Almeida and Torres, 2012a). With our choice of approximation, the fractional problem is transformed into a classical (integer-order) boundary value problem:

$$\begin{cases} \dot{x}(t) = \dfrac{\left(\frac{\alpha+2}{t} - At^{-\alpha}\right)x(t) + \sum_{p=2}^N C_p t^{1-p-\alpha} V_p(t) - \frac{\lambda(t)}{2t^2} + t^2}{1 + Bt^{1-\alpha}} \\ \dot{V}_p(t) = (1-p)t^{p-2}x(t), \quad p = 2, \ldots, N \\ \dot{\lambda}(t) = \dfrac{\left(A(1-t)^{-\alpha} - \frac{\alpha+2}{t}\right)\lambda(t) - \sum_{p=2}^N C_p(1-t)^{1-p-\alpha}W_p(t)}{1 + B(1-t)^{1-\alpha}} \\ \dot{W}_p(t) = -(1-p)(1-t)^{p-2}\lambda(t), \quad p = 2, \ldots, N, \end{cases}$$

subject to the boundary conditions

$$\begin{cases} x(0) = 0, \quad x(1) = \dfrac{2}{\Gamma(3+\alpha)}, \\ V_p(0) = 0, \quad p = 2, \ldots, N, \\ W_p(1) = 0, \quad p = 2, \ldots, N. \end{cases}$$

The solutions are depicted in Fig. 10.1 for $N = 2$, $N = 3$ and $\alpha = 1/2$. MATLAB® code can be found in the Appendix. Since the exact solution for this problem is known, for each N we compute the approximation error by using the maximum norm. Assume that $\bar{x}(t_i)$ are the approximated values on the discrete-time horizon $a = t_0, t_1, \ldots, t_n$. Then the error is given by

$$E = \max_i(|x(t_i) - \bar{x}(t_i)|).$$

Another approach is to approximate the original problem by using (6.10) for the fractional derivative. Following the procedure discussed in Section 10.1, the problem of Example 10.1 is approximated by

$$\tilde{J}[x, u] = \int_0^1 (tu - (\alpha + 2)x)^2 \, dt \longrightarrow \min$$

subject to the control system

$$\begin{cases} \dot{x}(t)[1 + B(\alpha, N)t^{1-\alpha}] + A(\alpha, N)t^{-\alpha}x(t) - \sum_{p=2}^N C(\alpha, p)t^{1-p-\alpha}V_p(t) \\ \qquad = u(t) + t^2 \\ \dot{V}_p(t) = (1-p)t^{p-2}x(t), \end{cases}$$

and boundary conditions

$$x(0) = 0, \quad x(1) = \frac{2}{\Gamma(3+\alpha)}, \quad V_p(0) = 0, \quad p = 2, 3, \ldots, N.$$

The Hamiltonian system for this classical optimal control problem is

$$H = \frac{\lambda_1(-A(\alpha, N)t^{-\alpha}x + \sum_{p=2}^N C(\alpha, p)t^{1-p-\alpha}V_p + u + t^2)}{1 + B(\alpha, N)t^{1-\alpha}}$$

$$+ \sum_{p=2}^N (1-p)t^{p-2}\lambda_p x + (tu - (\alpha + 2)x)^2.$$

Using the stationary condition $\frac{\partial H}{\partial u} = 0$, we have

$$u(t) = \frac{\alpha + 2}{t}x(t) - \frac{\lambda_1(t)}{2t^2(1 + B(\alpha, N)t^{1-\alpha})} \quad \text{for } t \neq 0.$$

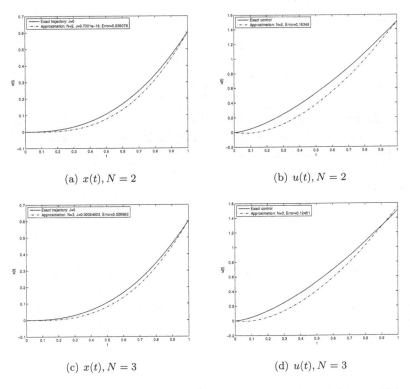

(a) $x(t)$, $N = 2$

(b) $u(t)$, $N = 2$

(c) $x(t)$, $N = 3$

(d) $u(t)$, $N = 3$

Fig. 10.1　Exact solution (solid lines) for the problem in Example 10.1 with $\alpha = 1/2$ *versus* numerical solutions (dashed lines) obtained using approximations (6.10) and (6.14) up to order N in the fractional necessary optimality conditions.

Finally, the Hamiltonian becomes

$$H = \phi_0\lambda_1^2 + \phi_1 x\lambda_1 + \sum_{p=2}^{N}\phi_p V_p\lambda_1 + \phi_{N+1}\lambda_1 + \sum_{p=2}^{N}(1-p)t^{p-2}x\lambda_p, \quad t \neq 0,$$

$$(10.12)$$

where

$$\phi_0(t) = \frac{-1}{4t^2(1 + B(\alpha, N)t^{1-\alpha})^2}, \quad \phi_1(t) = \frac{\alpha + 2 - A(\alpha, N)t^{1-\alpha}}{t(1 + B(\alpha, N)t^{1-\alpha})}, \quad (10.13)$$

and

$$\phi_p(t) = \frac{C(\alpha, p)t^{1-p-\alpha}}{1 + B(\alpha, N)t^{1-\alpha}}, \quad \phi_{N+1}(t) = \frac{t^2}{1 + B(\alpha, N)t^{1-\alpha}}. \quad (10.14)$$

The Hamiltonian system $\dot{\mathbf{x}} = \frac{\partial H}{\partial \boldsymbol{\lambda}}$, $\dot{\boldsymbol{\lambda}} = -\frac{\partial H}{\partial \mathbf{x}}$, gives

$$\begin{cases} \dot{x}(t) = 2\phi_0(t)\lambda_1(t) + \phi_1(t)x(t) + \sum_{p=2}^{N}\phi_p(t)V_p(t) + \phi_{N+1}(t) \\ \dot{V}_p(t) = (1-p)t^{p-2}x(t), \quad p = 2,\dots,N \\ \dot{\lambda}_1(t) = -\phi_1(t)\lambda_1(t) + \sum_{p=2}^{N}(p-1)t^{p-2}\lambda_p(t) \\ \dot{\lambda}_p(t) = -\phi_p(t)\lambda_1(t), \quad p = 2,\dots,N, \end{cases}$$

subject to the boundary conditions

$$\begin{cases} x(0) = 0 \\ V_p(0) = 0, \quad p = 2,\dots,N, \end{cases} \qquad \begin{cases} x(1) = \dfrac{2}{\Gamma(3+\alpha)} \\ \lambda_p(1) = 0, \quad p = 2,\dots,N. \end{cases}$$

This two-point boundary value problem was solved using MATLAB® bvp4c built-in function for $N = 2$ and $N = 3$. The results are depicted in Fig. 10.2.

10.4.2 Free final time

The two numerical methods discussed in Section 10.4.1 are now employed to solve a fractional-order optimal control problem with free final time T.

Example 10.2. Find an optimal triplet (x, u, T) that minimizes

$$J[x, u, T] = \int_0^T (tu(t) - (\alpha + 2)x(t))^2 \, dt$$

subject to the control system

$$\dot{x}(t) + {}_0^C D_t^\alpha x(t) = u(t) + t^2$$

and boundary conditions

$$x(0) = 0, \quad x(T) = 1.$$

An exact solution to this problem is not known and we apply the two numerical procedures already used with respect to the fixed final time problem in Example 10.1.

We begin by using the fractional necessary optimality conditions that, after approximating the fractional terms, results in

$$\begin{cases} \dot{x}(t) = \dfrac{\left(\frac{\alpha+2}{t} - At^{-\alpha}\right)x(t) + \sum_{p=2}^{N}C_p t^{1-p-\alpha}V_p(t) - \frac{\lambda(t)}{2t^2} + t^2}{1 + Bt^{1-\alpha}} \\ \dot{V}_p(t) = (1-p)t^{p-2}x(t), \quad p = 2,\dots,N \\ \dot{\lambda}(t) = \dfrac{\left(A(1-t)^{-\alpha} - \frac{\alpha+2}{t}\right)\lambda(t) - \sum_{p=2}^{N}C_p(1-t)^{1-p-\alpha}W_p(t)}{1 + B(1-t)^{1-\alpha}} \\ \dot{W}_p(t) = -(1-p)(1-t)^{p-2}\lambda(t), \quad p = 2,\dots,N, \end{cases}$$

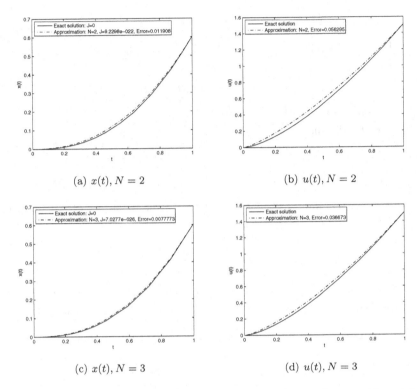

(a) $x(t), N = 2$ (b) $u(t), N = 2$

(c) $x(t), N = 3$ (d) $u(t), N = 3$

Fig. 10.2 Exact solution (solid lines) for the problem in Example 10.1 with $\alpha = 1/2$ *versus* numerical solutions (dashed lines) obtained by approximating the fractional-order optimal control problem using (6.10) up to order N and then solving the classical necessary optimality conditions with MATLAB® **bvp4c** built-in function.

subject to the boundary conditions

$$\begin{cases} x(0) = 0, \quad x(T) = 1, \\ V_p(0) = 0, \quad p = 2, \ldots, N, \\ W_p(T) = 0, \quad p = 2, \ldots, N. \end{cases}$$

The only difference here with respect to Example 10.1 is that there is an extra unknown, the terminal time T. The boundary condition for this new unknown is chosen appropriately from the transversality conditions discussed in Corollary 10.1, i.e.,

$$[H(t, x, u, \lambda) - \lambda(t)\,{}_a^C D_t^\alpha x(t) + \dot{x}(t)\,{}_t I_T^{1-\alpha}\lambda(t)]_{t=T} = 0,$$

where H is given as in (10.11). Since we require λ to be continuous, ${}_t I_T^{1-\alpha}\lambda(t)|_{t=T} = 0$ (cf. (Miller and Ross, 1993, p. 46)) and so $\lambda(T) = 0$.

One possible way to proceed consists in translating the problem into the interval $[0, 1]$ by the change of variable $t = Ts$ (Avvakumov and Kiselev, 2000). In this setting, either we add T to the problem as a new state variable with dynamics $\dot{T}(s) = 0$, or we treat it as a parameter. We use the latter, to get the following parametric boundary value problem:

$$
\begin{cases}
\dot{x}(s) = \dfrac{T}{1 + B(Ts)^{1-\alpha}} \\
\quad \times \left[\left(\frac{\alpha+2}{Ts} - A(Ts)^{-\alpha} \right) x(s) + \sum_{p=2}^{N} C_p (Ts)^{1-p-\alpha} V_p(s) - \frac{\lambda(s)}{2(Ts)^2} + (Ts)^2 \right], \\
\dot{V}_p(s) = T(1-p)(Ts)^{p-2} x(s), \quad p = 2, \dots, N, \\
\dot{\lambda}(s) = \dfrac{\left[\left(A(1 - Ts)^{-\alpha} - \frac{\alpha+2}{Ts} \right) \lambda(s) - \sum_{p=2}^{N} C_p (1 - Ts)^{1-p-\alpha} W_p(s) \right] T}{1 + B(1 - Ts)^{1-\alpha}}, \\
\dot{W}_p(s) = -T(1-p)(1 - Ts)^{p-2} \lambda(s), \quad p = 2, \dots, N,
\end{cases}
$$

subject to the boundary conditions

$$
\begin{cases}
x(0) = 0 \\
V_p(0) = 0, \quad p = 2, \dots, N \\
W_p(1) = 0, \quad p = 2, \dots, N,
\end{cases}
\qquad
\begin{cases}
x(1) = 1 \\
\lambda(1) = 0.
\end{cases}
$$

This parametric boundary value problem is solved for $N = 2$ and $\alpha = 0.5$ with MATLAB® bvp5c function (see Appendix). The result is shown in Fig. 10.3 (dashed lines).

We also solve Example 10.2 with $\alpha = 1/2$ by directly transforming it into an integer-order optimal control problem with free final time. As is well known in the classical theory of optimal control, the Hamiltonian must vanish at the terminal point when the final time is free, i.e., one has $H|_{t=T} = 0$ with H given by (10.12) (Kirk, 1970). For $N = 2$, the necessary optimality conditions give the following two point boundary value problem:

$$
\begin{cases}
\dot{x}(t) = 2\phi_0(t)\lambda_1(t) + \phi_1(t)x(t) + \phi_2(t)V_2(t) + \phi_3(t) \\
\dot{V}_2(t) = -x(t) \\
\dot{\lambda}_1(t) = -\phi_1(t)\lambda_1(t) + x(t) \\
\dot{\lambda}_2(t) = -\phi_2(t)\lambda_1(t),
\end{cases}
$$

where $\phi_0(t)$ and $\phi_1(t)$ are given by (10.13) and $\phi_2(t)$ and $\phi_3(t)$ by (10.14) with $p = N = 2$. The trajectory x and corresponding u are shown in Fig. 10.3 (dash-dotted lines).

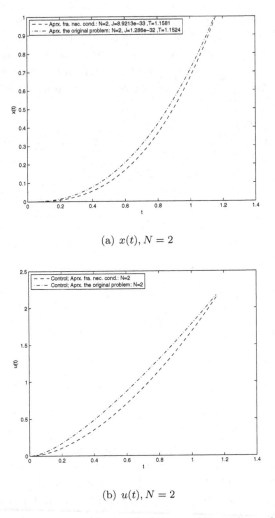

(a) $x(t), N = 2$

(b) $u(t), N = 2$

Fig. 10.3   Numerical solutions to the free final time problem of Example 10.2 with $\alpha = 1/2$, using fractional necessary optimality conditions (dashed lines) and approximation of the problem to an integer-order optimal control problem (dash-dotted lines).

Chapter 11

An expansion formula for fractional operators of variable order

We now obtain approximation formulas for fractional integrals and derivatives of Riemann–Liouville and Marchaud types with a variable fractional order. The approximations involve integer-order derivatives only. An estimation for the error is given. The efficiency of the approximation method is illustrated with examples. As applications, we show how the obtained results are useful to solve differential equations, and problems of the calculus of variations that depend on fractional derivatives of Marchaud type.

11.1 Introduction

Since the order α of the integrals and derivatives may take any value, an interesting extension is to consider the order not to be a constant during the process, but a variable $\alpha(t)$ that depends on time. This provides an extension of the classical fractional calculus and was introduced by Samko and Ross in 1993 (Samko and Ross, 1993) (see also (Samko, 1995)). The variable order fractional calculus is nowadays recognized as a useful tool, with successful applications in mechanics, in the modeling of linear and nonlinear visco-elasticity oscillators, and other phenomena where the order of the derivative varies with time. For more on the subject, and applications of it, we mention (Almeida and Samko, 2009; Coimbra, 2003, 2005; Diaz and Coimbra, 2009; Lorenzo and Hartley, 2002; Ramirez and Coimbra, 2010, 2011). For a numerical approach see, e.g., (Ma, Xu and Yue, 2012; Valério and Costa, 2011; Zhuang et al., 2009). Results on differential equations and the calculus of variations with fractional operators of variable order can be found in (Odzijewicz, Malinowska and Torres, 2013a,c) and references therein. Here we show how fractional derivatives and integrals of variable order can be approximated by classical integer-order operators.

The outline of this chapter follows (Almeida and Torres, 2013). In Section 11.2 we present the necessary definitions, namely the fractional operators of Riemann–Liouville and Marchaud of variable order. Some properties of the operators are also given. The main core of this chapter is Section 11.3, where we prove the expansion formulas for the considered fractional operators, with the size of the expansion being the derivative of order $n \in \mathbb{N}$. In Section 11.4 we show the accuracy of the method with some examples and how the approximations can be applied in different situations to solve problems involving variable order fractional operators.

11.2 Fractional calculus of variable order

In the following, the order of the fractional operators is given by a function $\alpha \in C^1([a, b],]0, 1[)$; x is assumed to ensure convergence for each of the involved integrals. For a complete and rigorous study of fractional calculus we refer to (Samko, Kilbas and Marichev, 1993).

Definition 11.1. *Let x be a function with domain $[a, b]$. Then,*

- *the left Riemann–Liouville fractional integral of order $\alpha(t)$ is given by*

$$_aI_t^{\alpha(t)}x(t) = \frac{1}{\Gamma(\alpha(t))} \int_a^t (t - \tau)^{\alpha(t)-1}x(\tau)d\tau,$$

- *the right Riemann–Liouville fractional integral of order $\alpha(t)$ is given by*

$$_tI_b^{\alpha(t)}x(t) = \frac{1}{\Gamma(\alpha(t))} \int_t^b (\tau - t)^{\alpha(t)-1}x(\tau)d\tau,$$

- *the left Riemann–Liouville fractional derivative of order $\alpha(t)$ is given by*

$$_aD_t^{\alpha(t)}x(t) = \frac{1}{\Gamma(1 - \alpha(t))} \frac{d}{dt} \int_a^t (t - \tau)^{-\alpha(t)}x(\tau)d\tau,$$

- *the right Riemann–Liouville fractional derivative of order $\alpha(t)$ is given by*

$$_tD_b^{\alpha(t)}x(t) = \frac{-1}{\Gamma(1 - \alpha(t))} \frac{d}{dt} \int_t^b (\tau - t)^{-\alpha(t)}x(\tau)d\tau,$$

- *the left Marchaud fractional derivative of order $\alpha(t)$ is given by*

$$_a\mathbb{D}_t^{\alpha(t)}x(t) = \frac{x(t)}{\Gamma(1 - \alpha(t))(t - a)^{\alpha(t)}} + \frac{\alpha(t)}{\Gamma(1 - \alpha(t))} \int_a^t \frac{x(t) - x(\tau)}{(t - \tau)^{1+\alpha(t)}}d\tau,$$
$$\tag{11.1}$$

- *the right Marchaud fractional derivative of order $\alpha(t)$ is given by*

$$_t\mathbb{D}_b^{\alpha(t)}x(t) = \frac{x(t)}{\Gamma(1-\alpha(t))(b-t)^{\alpha(t)}} + \frac{\alpha(t)}{\Gamma(1-\alpha(t))}\int_t^b \frac{x(t)-x(\tau)}{(\tau-t)^{1+\alpha(t)}}d\tau.$$

Remark 11.1. It follows from Definition 11.1 that

$$_aD_t^{\alpha(t)}x(t) = \frac{d}{dt}{_aI_t^{1-\alpha(t)}}x(t) \quad \text{and} \quad _tD_b^{\alpha(t)}x(t) = -\frac{d}{dt}{_tI_b^{1-\alpha(t)}}x(t).$$

Example 11.1 (See (Samko and Ross, 1993)). *Let x be the power function $x(t) = (t-a)^\gamma$. Then, for $\gamma > -1$, we have*

$$_aI_t^{\alpha(t)}x(t) = \frac{\Gamma(\gamma+1)}{\Gamma(\gamma+\alpha(t)+1)}(t-a)^{\gamma+\alpha(t)},$$

$$_a\mathbb{D}_t^{\alpha(t)}x(t) = \frac{\Gamma(\gamma+1)}{\Gamma(\gamma-\alpha(t)+1)}(t-a)^{\gamma-\alpha(t)},$$

and

$$_aD_t^{\alpha(t)}x(t) = \frac{\Gamma(\gamma+1)}{\Gamma(\gamma-\alpha(t)+1)}(t-a)^{\gamma-\alpha(t)}$$

$$-\frac{\alpha^{(1)}(t)\Gamma(\gamma+1)}{\Gamma(\gamma-\alpha(t)+2)}(t-a)^{\gamma-\alpha(t)+1}\left[\ln(t-a)-\psi(\gamma-\alpha(t)+2)+\psi(1-\alpha(t))\right],$$

where ψ is the Psi function, that is, the derivative of the logarithm of the Gamma function:

$$\psi(t) = \frac{d}{dt}\ln(\Gamma(t)) = \frac{\Gamma'(t)}{\Gamma(t)}.$$

From Example 11.1 we see that $_aD_t^{\alpha(t)}x(t) \neq {_a\mathbb{D}_t^{\alpha(t)}}x(t)$. Also, the symmetry on power functions is violated when we consider $_aI_t^{\alpha(t)}x(t)$ and $_aD_t^{\alpha(t)}x(t)$, but holds for $_aI_t^{\alpha(t)}x(t)$ and $_a\mathbb{D}_t^{\alpha(t)}x(t)$. Later we explain this better, when we deduce the expansion formula for the Marchaud fractional derivative. In contrast with the constant fractional-order case, the law of exponents fails for fractional integrals of variable order. However, a weak form holds (see (Samko and Ross, 1993)): if $\beta(t) \equiv \beta$, $t \in]0,1[$, then $_aI_t^{\alpha(t)}{_aI_t^{\beta}}x(t) = {_aI_t^{\alpha(t)+\beta}}x(t)$.

11.3 Expansion formulas with higher-order derivatives

The main results of this chapter provide approximations of the fractional derivatives of a given function x by sums involving only integer derivatives of x. The approximations use the generalization of the binomial coefficient formula to real numbers:

$$\binom{-\alpha(t)}{k}(-1)^k = \binom{\alpha(t)+k-1}{k} = \frac{\Gamma(\alpha(t)+k)}{\Gamma(\alpha(t))k!}.$$

Theorem 11.1. *Fix $n \in \mathbb{N}$ and $N \geq n+1$, and let $x \in C^{n+1}([a,b], \mathbb{R})$. Define the (left) moment of x of order k by*

$$V_k(t) = (k-n)\int_a^t (\tau-a)^{k-n-1}x(\tau)d\tau.$$

Then,

$$_aD_t^{\alpha(t)}x(t) = S_1(t) - S_2(t) + E_{1,N}(t) + E_{2,N}(t)$$

with

$$S_1(t) = (t-a)^{-\alpha(t)}\left[\sum_{k=0}^n A(\alpha(t),k)(t-a)^k x^{(k)}(t)\right.$$

$$\left. + \sum_{k=n+1}^N B(\alpha(t),k)(t-a)^{n-k}V_k(t)\right], \quad (11.2)$$

where

$$A(\alpha(t),k) = \frac{1}{\Gamma(k+1-\alpha(t))}\left[1 + \sum_{p=n+1-k}^N \frac{\Gamma(p-n+\alpha(t))}{\Gamma(\alpha(t)-k)(p-n+k)!}\right],$$

$$B(\alpha(t),k) = \frac{\Gamma(k-n+\alpha(t))}{\Gamma(-\alpha(t))\Gamma(1+\alpha(t))(k-n)!},$$

$k = 0,\ldots,n,$ *and*

$$S_2(t) = \frac{x(t)\alpha^{(1)}(t)}{\Gamma(1-\alpha(t))}(t-a)^{1-\alpha(t)}\left[\frac{\ln(t-a)}{1-\alpha(t)} - \frac{1}{(1-\alpha(t))^2}\right.$$

$$\left. - \ln(t-a)\sum_{k=0}^N \binom{-\alpha(t)}{k}\frac{(-1)^k}{k+1} + \sum_{k=0}^N \binom{-\alpha(t)}{k}(-1)^k\sum_{p=1}^N \frac{1}{p(k+p+1)}\right]$$

$$+ \frac{\alpha^{(1)}(t)(t-a)^{1-\alpha(t)}}{\Gamma(1-\alpha(t))}\left[\ln(t-a)\sum_{k=n+1}^{N+n+1}\binom{-\alpha(t)}{k-n-1}\frac{(-1)^{k-n-1}}{k-n}(t-a)^{n-k}V_k(t)\right.$$

$$\left. - \sum_{k=n+1}^{N+n+1}\binom{-\alpha(t)}{k-n-1}(-1)^{k-n-1}\sum_{p=1}^N \frac{1}{p(k+p-n)}(t-a)^{n-k-p}V_{k+p}(t)\right].$$

$$(11.3)$$

The error of the approximation $_aD_t^{\alpha(t)}x(t) \approx S_1(t) - S_2(t)$ is given by $E_{1,N}(t) + E_{2,N}(t)$, where $E_{1,N}(t)$ and $E_{2,N}(t)$ are bounded by

$$|E_{1,N}(t)| \leq L_{n+1}(t)\frac{\exp((n-\alpha(t))^2 + n - \alpha(t))}{\Gamma(n+1-\alpha(t))(n-\alpha(t))N^{n-\alpha(t)}}(t-a)^{n+1-\alpha(t)}$$

$$(11.4)$$

and

$$|E_{2,N}(t)| \leq \frac{L_1(t)\left|\alpha^{(1)}(t)\right|(t-a)^{2-\alpha(t)}\exp(\alpha^2(t)-\alpha(t))}{\Gamma(2-\alpha(t))N^{1-\alpha(t)}}\left[|\ln(t-a)| + \frac{1}{N}\right]$$

$$(11.5)$$

with

$$L_j(t) = \max_{\tau \in [a,t]}\left|x^{(j)}(\tau)\right|, \quad j \in \{1, n+1\}.$$

Proof. Starting with equality

$$_aD_t^{\alpha(t)}x(t) = \frac{1}{\Gamma(1-\alpha(t))}\frac{d}{dt}\int_a^t (t-\tau)^{-\alpha(t)}x(\tau)d\tau,$$

doing the change of variable $t - \tau = u - a$ over the integral, and then differentiating it, we get

$$_aD_t^{\alpha(t)}x(t) = \frac{1}{\Gamma(1-\alpha(t))}\frac{d}{dt}\int_a^t (u-a)^{-\alpha(t)}x(t-u+a)du$$

$$= \frac{1}{\Gamma(1-\alpha(t))}\left[\frac{x(a)}{(t-a)^{\alpha(t)}} + \int_a^t \frac{d}{dt}\left[(u-a)^{-\alpha(t)}x(t-u+a)\right]du\right]$$

$$= \frac{1}{\Gamma(1-\alpha(t))}\left[\frac{x(a)}{(t-a)^{\alpha(t)}} + \int_a^t \left[-\alpha^{(1)}(t)(u-a)^{-\alpha(t)}\ln(u-a)\right.\right.$$

$$\left.\left. \times\, x(t-u+a) + (u-a)^{-\alpha(t)}x^{(1)}(t-u+a)\right]du\right]$$

$$= S_1(t) - S_2(t)$$

with

$$S_1(t) = \frac{1}{\Gamma(1-\alpha(t))}\left[\frac{x(a)}{(t-a)^{\alpha(t)}} + \int_a^t (t-\tau)^{-\alpha(t)}x^{(1)}(\tau)d\tau\right] \quad (11.6)$$

and

$$S_2(t) = \frac{\alpha^{(1)}(t)}{\Gamma(1-\alpha(t))}\int_a^t (t-\tau)^{-\alpha(t)}\ln(t-\tau)x(\tau)d\tau. \quad (11.7)$$

The equivalence between (11.6) and (11.2) follows from the computations of (Pooseh, Almeida and Torres, 2013a). To show the equivalence between

(11.7) and (11.3) we start in the same way as shown in (Atanacković *et al.*, 2013), to get

$$
S_2(t) = \frac{\alpha^{(1)}(t)}{\Gamma(1-\alpha(t))} \left[x(t) \int_a^t (t-u)^{-\alpha(t)} \ln(t-u)\, du \right.
$$
$$
\left. - \int_a^t x^{(1)}(\tau) \left(\int_a^\tau (t-u)^{-\alpha(t)} \ln(t-u)\, du \right) d\tau \right]
$$
$$
= \frac{\alpha^{(1)}(t)}{\Gamma(1-\alpha(t))} \left[x(t)(t-a)^{1-\alpha(t)} \left[\frac{\ln(t-a)}{1-\alpha(t)} - \frac{1}{(1-\alpha(t))^2} \right] \right.
$$
$$
- \int_a^t x^{(1)}(\tau) \left(\int_a^\tau (t-a)^{-\alpha(t)} \left(1 - \frac{u-a}{t-a} \right)^{-\alpha(t)} \right.
$$
$$
\left. \left. \times \left[\ln(t-a) + \ln\left(1 - \frac{u-a}{t-a} \right) \right] du \right) d\tau \right].
$$

Now, applying Taylor's expansion over

$$
\left(1 - \frac{u-a}{t-a} \right)^{-\alpha(t)} \quad \text{and} \quad \ln\left(1 - \frac{u-a}{t-a} \right),
$$

we deduce that

$$
S_2(t) = \frac{\alpha^{(1)}(t)}{\Gamma(1-\alpha(t))} \left[x(t)(t-a)^{1-\alpha(t)} \left[\frac{\ln(t-a)}{1-\alpha(t)} - \frac{1}{(1-\alpha(t))^2} \right] \right.
$$
$$
- \int_a^t x^{(1)}(\tau) \left(\int_a^\tau (t-a)^{-\alpha(t)} \ln(t-a) \sum_{k=0}^N \binom{-\alpha(t)}{k} (-1)^k \frac{(u-a)^k}{(t-a)^k}\, du \right.
$$
$$
\left. \left. - \int_a^\tau (t-a)^{-\alpha(t)} \sum_{k=0}^N \binom{-\alpha(t)}{k} (-1)^k \frac{(u-a)^k}{(t-a)^k} \sum_{p=1}^N \frac{1}{p}\frac{(u-a)^p}{(t-a)^p}\, du \right) d\tau \right]
$$
$$
+ E_{2,N}(t)
$$
$$
= \frac{\alpha^{(1)}(t)}{\Gamma(1-\alpha(t))} \left[x(t)(t-a)^{1-\alpha(t)} \left[\frac{\ln(t-a)}{1-\alpha(t)} - \frac{1}{(1-\alpha(t))^2} \right] \right.
$$
$$
- \int_a^t x^{(1)}(\tau)(t-a)^{-\alpha(t)} \ln(t-a)
$$
$$
\times \sum_{k=0}^N \binom{-\alpha(t)}{k} \frac{(-1)^k}{(t-a)^k} \left(\int_a^\tau (u-a)^k\, du \right) d\tau + \int_a^t x^{(1)}(\tau)(t-a)^{-\alpha(t)}
$$
$$
\left. \times \sum_{k=0}^N \binom{-\alpha(t)}{k} \frac{(-1)^k}{(t-a)^k} \sum_{p=1}^N \frac{1}{p(t-a)^p} \left(\int_a^\tau (u-a)^{k+p}\, du \right) d\tau \right]
$$
$$
+ E_{2,N}(t)
$$

$$= \frac{\alpha^{(1)}(t)(t-a)^{-\alpha(t)}}{\Gamma(1-\alpha(t))} \left[x(t)(t-a) \left[\frac{\ln(t-a)}{1-\alpha(t)} - \frac{1}{(1-\alpha(t))^2} \right] \right.$$

$$- \ln(t-a) \sum_{k=0}^{N} \binom{-\alpha(t)}{k} \frac{(-1)^k}{(t-a)^k(k+1)} \left(\int_a^t x^{(1)}(\tau)(\tau-a)^{k+1}\, d\tau \right)$$

$$+ \sum_{k=0}^{N} \binom{-\alpha(t)}{k} \frac{(-1)^k}{(t-a)^k} \sum_{p=1}^{N} \frac{1}{p(t-a)^p(k+p+1)}$$

$$\left. \times \left(\int_a^t x^{(1)}(\tau)(\tau-a)^{k+p+1}\, d\tau \right) \right] + E_{2,N}(t).$$

Integrating by parts, we conclude with the two following equalities:

$$\int_a^t x^{(1)}(\tau)(\tau-a)^{k+1}\, d\tau = x(t)(t-a)^{k+1} - V_{k+n+1}(t),$$

$$\int_a^t x^{(1)}(\tau)(\tau-a)^{k+p+1}\, d\tau = x(t)(t-a)^{k+p+1} - V_{k+p+n+1}(t).$$

The deduction of relation (11.3) for $S_2(t)$ follows now from direct calculations. To end, we prove the upper bound formula for the error. The bound (11.4) for the error $E_{1,N}(t)$ at time t follows easily from (Pooseh, Almeida and Torres, 2013a). With respect to sum S_2, the error at t is bounded by

$$|E_{2,N}(t)| \leq \left| \frac{\alpha^{(1)}(t)(t-a)^{-\alpha(t)}}{\Gamma(1-\alpha(t))} \right|$$

$$\times \left| -\ln(t-a) \sum_{k=N+1}^{\infty} \binom{-\alpha(t)}{k} \frac{(-1)^k}{k+1} \left(\int_a^t x^{(1)}(\tau) \frac{(\tau-a)^{k+1}}{(t-a)^k}\, d\tau \right) \right.$$

$$\left. + \sum_{k=N+1}^{\infty} \binom{-\alpha(t)}{k} (-1)^k \sum_{p=N+1}^{\infty} \frac{1}{p(k+p+1)} \left(\int_a^t x^{(1)}(\tau) \frac{(\tau-a)^{k+p+1}}{(t-a)^{k+p}}\, d\tau \right) \right|.$$

Define the quantities

$$I_1(t) = \int_N^{\infty} \frac{1}{k^{1-\alpha(t)}(k+1)(k+2)}\, dk$$

and

$$I_2(t) = \int_N^{\infty} \int_N^{\infty} \frac{1}{k^{1-\alpha(t)}p(k+p+1)(k+p+2)}\, dp\, dk.$$

Inequality (11.5) follows from relation

$$\left| \binom{-\alpha(t)}{k} \right| \leq \frac{\exp(\alpha^2(t)-\alpha(t))}{k^{1-\alpha(t)}}.$$

and the upper bounds

$$I_1(t) < \int_N^\infty \frac{1}{k^{2-\alpha(t)}}\, dk = \frac{1}{(1-\alpha(t))N^{1-\alpha(t)}}$$

and

$$I_2(t) < \int_N^\infty \int_N^\infty \frac{1}{k^{2-\alpha(t)}p^2}\, dp\, dk = \frac{1}{(1-\alpha(t))N^{2-\alpha(t)}}$$

for I_1 and I_2.   □

Similarly as concluded in Theorem 11.1 for the left Riemann–Liouville fractional derivative, an approximation formula can be deduced for the right Riemann–Liouville fractional derivative:

Theorem 11.2. *Fix $n \in \mathbb{N}$ and $N \geq n+1$, and let $x \in C^{n+1}([a,b], \mathbb{R})$. Define the (right) moment of x of order k by*

$$W_k(t) = (k-n) \int_t^b (b-\tau)^{k-n-1} x(\tau) d\tau.$$

Then,

$$_tD_b^{\alpha(t)} x(t) = S_1(t) + S_2(t) + E_{1,N}(t) + E_{2,N}(t)$$

with

$$S_1(t) = (b-t)^{-\alpha(t)} \left[\sum_{k=0}^n A(\alpha(t),k)(b-t)^k x^{(k)}(t) \right.$$

$$\left. + \sum_{k=n+1}^N B(\alpha(t),k)(b-t)^{n-k} W_k(t) \right],$$

where

$$A(\alpha(t),k) = \frac{(-1)^k}{\Gamma(k+1-\alpha(t))} \left[1 + \sum_{p=n+1-k}^N \frac{\Gamma(p-n+\alpha(t))}{\Gamma(\alpha(t)-k)(p-n+k)!} \right],$$

$$B(\alpha(t),k) = \frac{(-1)^{n+1}\Gamma(k-n+\alpha(t))}{\Gamma(-\alpha(t))\Gamma(1+\alpha(t))(k-n)!},$$

$k = 0, \dots, n$, and

$$
S_2(t) = \frac{x(t)\alpha^{(1)}(t)}{\Gamma(1 - \alpha(t))}(b - t)^{1-\alpha(t)} \left[\frac{\ln(b - t)}{1 - \alpha(t)} - \frac{1}{(1 - \alpha(t))^2} \right.
$$

$$
\left. - \ln(b - t) \sum_{k=0}^{N} \binom{-\alpha(t)}{k} \frac{(-1)^k}{k + 1} + \sum_{k=0}^{N} \binom{-\alpha(t)}{k} (-1)^k \sum_{p=1}^{N} \frac{1}{p(k + p + 1)} \right]
$$

$$
+ \frac{\alpha^{(1)}(t)(b - t)^{1-\alpha(t)}}{\Gamma(1 - \alpha(t))} \left[\ln(b - t) \sum_{k=n+1}^{N+n+1} \binom{-\alpha(t)}{k-n-1} \frac{(-1)^{k-n-1}}{k - n}(b - t)^{n-k} W_k(t) \right.
$$

$$
\left. - \sum_{k=n+1}^{N+n+1} \binom{-\alpha(t)}{k-n-1} (-1)^{k-n-1} \sum_{p=1}^{N} \frac{1}{p(k + p - n)}(b - t)^{n-k-p} W_{k+p}(t) \right].
$$

The error of the approximation $_tD_b^{\alpha(t)}x(t) \approx S_1(t) + S_2(t)$ is given by $E_{1,N}(t) + E_{2,N}(t)$, where $E_{1,N}(t)$ and $E_{2,N}(t)$ are bounded by

$$
|E_{1,N}(t)| \leq L_{n+1}(t) \frac{\exp((n - \alpha(t))^2 + n - \alpha(t))}{\Gamma(n + 1 - \alpha(t))(n - \alpha(t))N^{n-\alpha(t)}}(b - t)^{n+1-\alpha(t)}
$$

and

$$
|E_{2,N}(t)| \leq \frac{L_1(t) \left| \alpha^{(1)}(t) \right| (b - t)^{2-\alpha(t)} \exp(\alpha^2(t) - \alpha(t))}{\Gamma(2 - \alpha(t))N^{1-\alpha(t)}} \left[|\ln(b - t)| + \frac{1}{N} \right]
$$

with

$$
L_j(t) = \max_{\tau \in [a,t]} \left| x^{(j)}(\tau) \right|, \quad j \in \{1, n + 1\}.
$$

Using the techniques presented in (Pooseh, Almeida and Torres, 2012a), similar formulas as the ones given by Theorem 11.1 and Theorem 11.2 can be proved for the left and right Riemann–Liouville fractional integrals of order $\alpha(t)$. For example, for the left fractional integral one has the following result.

Theorem 11.3. *Fix $n \in \mathbb{N}$ and $N \geq n + 1$, and let $x \in C^{n+1}([a, b], \mathbb{R})$. Then,*

$$
_aI_t^{\alpha(t)}x(t) = (t - a)^{\alpha(t)} \left[\sum_{k=0}^{n} A(\alpha(t), k)(t - a)^k x^{(k)}(t) \right.
$$

$$
\left. + \sum_{k=n+1}^{N} B(\alpha(t), k)(t - a)^{n-k} V_k(t) \right] + E_N(t),
$$

where

$$A(\alpha(t), k) = \frac{1}{\Gamma(k+1+\alpha(t))} \left[1 + \sum_{p=n+1-k}^{N} \frac{\Gamma(p-n-\alpha(t))}{\Gamma(-\alpha(t)-k)(p-n+k)!} \right],$$

$$B(\alpha(t), k) = \frac{\Gamma(k-n-\alpha(t))}{\Gamma(\alpha(t))\Gamma(1-\alpha(t))(k-n)!},$$

$k = 0, \ldots, n,$ *and*

$$V_k(t) = (k-n) \int_a^t (\tau - a)^{k-n-1} x(\tau) d\tau, \quad k = n+1, \ldots$$

A bound for the error $E_N(t)$ *is given by*

$$|E_N(t)| \le L_{n+1}(t) \frac{\exp((n+\alpha(t))^2 + n + \alpha(t))}{\Gamma(n+1+\alpha(t))(n+\alpha(t))N^{n+\alpha(t)}} (t-a)^{n+1+\alpha(t)}.$$

We now focus our attention to the left Marchaud fractional derivative $_a\mathbb{D}_t^{\alpha(t)} x(t)$. Splitting the integral (11.1), we deduce that

$$_a\mathbb{D}_t^{\alpha(t)} x(t) = -\frac{\alpha(t)}{\Gamma(1-\alpha(t))} \int_a^t \frac{x(\tau)}{(t-\tau)^{1+\alpha(t)}} d\tau.$$

Integrating by parts,

$$_a\mathbb{D}_t^{\alpha(t)} x(t) = \frac{1}{\Gamma(1-\alpha(t))} \left[\frac{x(a)}{(t-a)^{\alpha(t)}} + \int_a^t (t-\tau)^{-\alpha(t)} x^{(1)}(\tau) d\tau \right], \quad (11.8)$$

which is a representation for the left Riemann–Liouville fractional derivative when the order is constant, that is, $\alpha(t) \equiv \alpha$ (Samko, Kilbas and Marichev, 1993, Lemma 2.12). For this reason, the Marchaud fractional derivative is more suitable as the inverse operation for the Riemann–Liouville fractional integral. With Eq. (11.8) and Theorem 11.1 in mind, it is not difficult to obtain the corresponding formula for $_a\mathbb{D}_t^{\alpha(t)} x(t)$.

Theorem 11.4. *Fix* $n \in \mathbb{N}$ *and* $N \ge n+1$, *and let* $x \in C^{n+1}([a,b],\mathbb{R})$. *Then,*

$$_a\mathbb{D}_t^{\alpha(t)} x(t) = S_1(t) + E_{1,N}(t),$$

where $S_1(t)$ *and* $E_{1,N}(t)$ *are as in Theorem 11.1.*

Similarly, having in consideration that

$$_t\mathbb{D}_b^{\alpha(t)} x(t) = \frac{1}{\Gamma(1-\alpha(t))} \left[\frac{x(b)}{(b-t)^{\alpha(t)}} - \int_t^b (\tau-t)^{-\alpha(t)} x^{(1)}(\tau) d\tau \right],$$

the following result holds.

Theorem 11.5. *Fix* $n \in \mathbb{N}$ *and* $N \ge n+1$, *and let* $x \in C^{n+1}([a,b],\mathbb{R})$. *Then,*

$$_t\mathbb{D}_b^{\alpha(t)} x(t) = S_1(t) + E_{1,N}(t),$$

where $S_1(t)$ *and* $E_{1,N}(t)$ *are as in Theorem 11.2.*

11.4   Examples

For illustrative purposes, we consider the left Riemann–Liouville fractional integral and the left Riemann–Liouville and Marchaud fractional derivatives of order $\alpha(t) = (t+1)/4$. Similar results as the ones presented here are easily obtained for the other fractional operators and for other functions $\alpha(t)$. All computations were carried out using the Computer Algebra System Maple®.

11.4.1   *Test function*

We test the accuracy of the approximations with an example.

Example 11.2. Let x be the function $x(t) = t^4$ with $t \in [0,1]$. Then, for $\alpha(t) = (t+1)/4$, it follows from Example 11.1 that

$$_0I_t^{\alpha(t)}x(t) = \frac{24}{\Gamma(\frac{t+21}{4})} t^{\frac{t+17}{4}}, \tag{11.9}$$

$$_0D_t^{\alpha(t)}x(t) = \frac{24t^{\frac{15-t}{4}}}{\Gamma(\frac{19-t}{4})} - \frac{6t^{\frac{19-t}{4}}}{\Gamma(\frac{23-t}{4})} \left[\ln(t) - \psi\left(\frac{23-t}{4}\right) + \psi\left(\frac{3-t}{4}\right)\right], \tag{11.10}$$

and

$$_0\mathbb{D}_t^{\alpha(t)}x(t) = \frac{24}{\Gamma(\frac{19-t}{4})} t^{\frac{15-t}{4}}. \tag{11.11}$$

In Figs. 11.1, 11.2 and 11.3 one can compare the exact expressions of the fractional operators of variable order (11.9), (11.10) and (11.11), respectively, with the approximations obtained from the results of Section 11.3 with $n = 2$ and $N \in \{3,5\}$. The error E is measured using the norm

$$E[f,g] = \sqrt{\int_0^1 (f(t) - g(t))^2 \, dt}. \tag{11.12}$$

11.4.2   *Fractional differential equations of variable order*

Consider the following fractional differential equation of variable order:

$$\begin{cases} _0\mathbb{D}_t^{\alpha(t)}x(t) + x(t) = \frac{1}{\Gamma(\frac{7-t}{4})} t^{\frac{3-t}{4}} + t, \\ x(0) = 0, \end{cases} \tag{11.13}$$

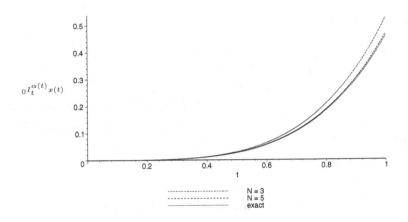

Fig. 11.1   Exact (11.9) and numerical approximations of the left Riemann–Liouville integral $_0I_t^{\alpha(t)}x(t)$ with $x(t) = t^4$ and $\alpha(t) = (t+1)/4$ obtained from Theorem 11.3 with $n = 2$ and $N \in \{3, 5\}$. The error (11.12) is $E \approx 0.02169$ for $N = 3$ and $E \approx 0.00292$ for $N = 5$.

with $\alpha(t) = (t+1)/4$. It is easy to check that $\bar{x}(t) = t$ is a solution to (11.13). We exemplify how Theorem 11.4 may be applied in order to approximate the solution of such type of problems. The main idea is to replace all the fractional operators that appear in the differential equation by a finite sum up to order N, involving integer derivatives only, and, by doing so, to obtain a new system of standard ordinary differential equations that is an approximation of the initial fractional variable order problem. As the size of N increases, the solution of the new system converges to the solution of the initial fractional system. The procedure for (11.13) is the following. First, we replace $_0\mathbb{D}_t^{\alpha(t)}x(t)$ by

$$_0\mathbb{D}_t^{\alpha(t)}x(t) \approx A(\alpha(t), N)t^{-\alpha(t)}x(t) + B(\alpha(t), N)t^{1-\alpha(t)}x^{(1)}(t)$$

$$+ \sum_{k=2}^{N} C(\alpha(t), k)t^{1-k-\alpha(t)}V_k(t),$$

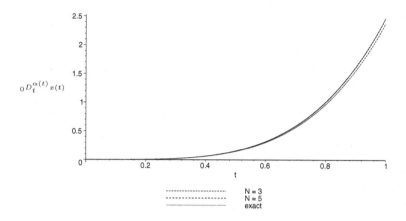

Fig. 11.2 Exact (11.10) and numerical approximations of the left Riemann–Liouville derivative $_0D_t^{\alpha(t)}x(t)$ with $x(t) = t^4$ and $\alpha(t) = (t+1)/4$ obtained from Theorem 11.1 with $n = 2$ and $N \in \{3, 5\}$. The error (11.12) is $E \approx 0.03294$ for $N = 3$ and $E \approx 0.003976$ for $N = 5$.

where

$$A(\alpha(t), N) = \frac{1}{\Gamma(1 - \alpha(t))} \left[1 + \sum_{p=2}^{N} \frac{\Gamma(p - 1 + \alpha(t))}{\Gamma(\alpha(t))(p - 1)!} \right],$$

$$B(\alpha(t), N) = \frac{1}{\Gamma(2 - \alpha(t))} \left[1 + \sum_{p=1}^{N} \frac{\Gamma(p - 1 + \alpha(t))}{\Gamma(\alpha(t) - 1)p!} \right],$$

$$C(\alpha(t), k) = \frac{\Gamma(k - 1 + \alpha(t))}{\Gamma(-\alpha(t))\Gamma(1 + \alpha(t))(k - 1)!},$$

and $V_k(t)$ is the solution of the system

$$\begin{cases} V_k^{(1)}(t) = (k - 1)t^{k-2}x(t) \\ V_k(0) = 0, \qquad k = 2, 3, \ldots, N. \end{cases}$$

Thus, we get the approximated system of ordinary differential equations

$$\begin{cases} \left[A(\alpha(t), N)t^{-\alpha(t)} + 1 \right] x(t) + B(\alpha(t), N)t^{1-\alpha(t)}x^{(1)}(t) \\ \quad + \sum_{k=2}^{N} C(\alpha(t), k)t^{1-k-\alpha(t)}V_k(t) = \frac{1}{\Gamma(\frac{7-t}{4})}t^{\frac{3-t}{4}} + t, \\ V_k^{(1)}(t) = (k - 1)t^{k-2}x(t), \qquad k = 2, 3, \ldots, N, \\ x(0) = 0, \\ V_k(0) = 0, \qquad k = 2, 3, \ldots, N. \end{cases} \qquad (11.14)$$

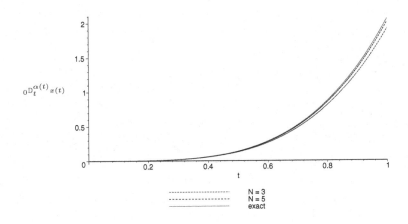

Fig. 11.3   Exact (11.11) and numerical approximations of the left Marchaud derivative $_0\mathbb{D}_t^{\alpha(t)}x(t)$ with $x(t) = t^4$ and $\alpha(t) = (t+1)/4$ obtained from Theorem 11.4 with $n = 2$ and $N \in \{3, 5\}$. The error (11.12) is $E \approx 0.04919$ for $N = 3$ and $E \approx 0.01477$ for $N = 5$.

Now we apply any standard technique to solve the system of ordinary differential equations (11.14). We used the command dsolve of Maple®. In Fig. 11.4 we find the graph of the approximation $\tilde{x}_3(t)$ to the solution of problem (11.13), obtained solving (11.14) with $N = 3$. Table 11.1 gives some numerical values of such approximation, illustrating numerically the fact that the approximation $\tilde{x}_3(t)$ is already very close to the exact solution $\overline{x}(t) = t$ of (11.13). In fact the plot of $\tilde{x}_3(t)$ in Fig. 11.4 is visually indistinguishable from the plot of $\overline{x}(t) = t$.

Table 11.1   Some numerical values of the solution $\tilde{x}_3(t)$ of (11.14) with $N = 3$, very close to the values of the solution $\overline{x}(t) = t$ of the fractional differential equation of variable order (11.13).

t	0.2	0.4	0.6	0.8	1
$\tilde{x}_3(t)$	0.200000020	0.400000040	0.600000094	0.800000026	1.00000016

11.4.3   *Fractional variational calculus of variable order*

We now exemplify how the expansions obtained in Section 11.3 are useful to approximate solutions of fractional problems of the calculus of variations

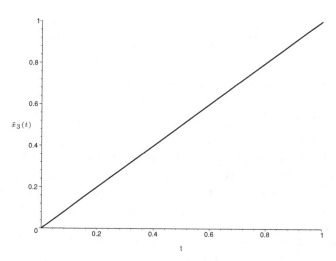

Fig. 11.4   Approximation $\tilde{x}_3(t)$ to the exact solution $\overline{x}(t) = t$ of the fractional differential equation (11.13), obtained from the application of Theorem 11.4, that is, obtained solving (11.14) with $N = 3$.

(Malinowska and Torres, 2012). The fractional variational calculus of variable order is a recent subject under strong current development (Odzijewicz, Malinowska and Torres, 2012c, 2013a,b,c). So far, only analytical methods to solve fractional problems of the calculus of variations of variable order have been developed in the literature, which consist in the solution of fractional Euler–Lagrange differential equations of variable order (Odzijewicz, Malinowska and Torres, 2012c, 2013a,b,c). In most cases, however, to solve analytically such fractional differential equations is extremely hard or even impossible, so numerical/approximating methods are needed. Our results provide two approaches to this issue. The first was already illustrated in Section 11.4.2 and consists in approximating the necessary optimality conditions proved in (Odzijewicz, Malinowska and Torres, 2012c, 2013a,b,c), which are nothing else than fractional differential equations of variable order.

The second approach is now considered. Similarly to Section 11.4.2, the main idea here is to replace the fractional operators of variable order that appear in the formulation of the variational problem by the corresponding expansion of Section 11.3, which involves only integer-order derivatives. By doing it, we reduce the original problem to a classical optimal control problem, whose extremals are found by applying the celebrated Pontryagin maximum principle (Pontryagin *et al.*, 1962). We illustrate this method with a concrete example. Consider the functional

$$J[x] = \int_0^1 \left[{}_0\mathbb{D}_t^{\alpha(t)} x(t) - \frac{1}{\Gamma(\frac{7-t}{4})} t^{\frac{3-t}{4}} \right]^2 dt, \qquad (11.15)$$

with fractional order $\alpha(t) = (t+1)/4$, subject to the boundary conditions

$$x(0) = 0, \quad x(1) = 1. \qquad (11.16)$$

Since $J[x] \geq 0$ for any admissible function x and taking $\overline{x}(t) = t$, which satisfies the given boundary conditions (11.16), one has $J[\overline{x}] = 0$, we conclude that $\overline{x}$ gives the global minimum to the fractional problem of the calculus of variations that consists in minimizing functional (11.15) subject to the boundary conditions (11.16). The numerical procedure is now explained. Since we have two boundary conditions, we replace ${}_0\mathbb{D}_t^{\alpha(t)} x(t)$ by the expansion given in Theorem 11.4 with $n = 1$ and a variable size $N \geq 2$. The approximation becomes

$$
{}_0\mathbb{D}_t^{\alpha(t)} x(t) \approx A(\alpha(t), N) t^{-\alpha(t)} x(t) + B(\alpha(t), N) t^{1-\alpha(t)} x^{(1)}(t)
$$
$$
+ \sum_{k=2}^N C(\alpha(t), k) t^{1-k-\alpha(t)} V_k(t). \qquad (11.17)
$$

Using (11.17), we approximate the initial problem by the following one: to minimize

$$
\tilde{J}[x] = \int_0^1 \left[A(\alpha(t), N) t^{-\alpha(t)} x(t) + B(\alpha(t), N) t^{1-\alpha(t)} x^{(1)}(t) \right.
$$
$$
\left. + \sum_{k=2}^N C(\alpha(t), k) t^{1-k-\alpha(t)} V_k(t) - \frac{1}{\Gamma(\frac{7-t}{4})} t^{\frac{3-t}{4}} \right]^2 dt
$$

subject to

$$V_k^{(1)}(t) = (k-1) t^{k-2} x(t), \quad V_k(0) = 0, \qquad k = 2, \ldots, N,$$

and

$$x(0) = 0, \quad x(1) = 1,$$

where $\alpha(t) = (t+1)/4$. This dynamic optimization problem has a system of ordinary differential equations as a constraint, so it is natural to solve it as an optimal control problem. For that, define the control u by

$$
u(t) = A(\alpha(t), N) t^{-\alpha(t)} x(t) + B(\alpha(t), N) t^{1-\alpha(t)} x^{(1)}(t)
$$
$$
+ \sum_{k=2}^N C(\alpha(t), k) t^{1-k-\alpha(t)} V_k(t).
$$

We then obtain the control system

$$x^{(1)}(t) = B^{-1}t^{\alpha(t)-1}u(t) - AB^{-1}t^{-1}x(t) - \underbrace{\sum_{k=2}^{N} B^{-1}C_k t^{-k}V_k(t)}_{}$$

$$\underbrace{\hspace{8cm}}_{f(t,x(t),u(t),V(t))}$$

where, for simplification,

$$A = A(\alpha(t), N), \quad B = B(\alpha(t), N), \quad C_k = C(\alpha(t), k)$$

and $V(t) = (V_2(t), \ldots, V_N(t))$. In conclusion, we wish to minimize the functional

$$\tilde{J}[x, u, V] = \int_0^1 \left[u(t) - \frac{1}{\Gamma(\frac{7-t}{4})} t^{\frac{3-t}{4}} \right]^2 dt$$

subject to the first-order dynamic constraints

$$\begin{cases} x^{(1)}(t) = f(t, x, u, V), \\ V_k^{(1)}(t) = (k-1)t^{k-2}x(t), \quad k = 2, \ldots, N, \end{cases}$$

and the boundary conditions

$$\begin{cases} x(0) = 0, \\ x(1) = 1, \\ V_k(0) = 0, \quad k = 2, \ldots, N. \end{cases}$$

In this case, the Hamiltonian is given by

$$H(t, x, u, V, \lambda) = \left[u - \frac{1}{\Gamma(\frac{7-t}{4})} t^{\frac{3-t}{4}} \right]^2 + \lambda_1 f(t, x, u, V) + \sum_{k=2}^{N} \lambda_k (k-1)t^{k-2}x$$

with the adjoint vector $\lambda = (\lambda_1, \lambda_2, \ldots, \lambda_N)$ (Pontryagin *et al.*, 1962). Following the classical optimal control approach of Pontryagin (Pontryagin *et al.*, 1962), we have the following necessary optimality conditions:

$$\frac{\partial H}{\partial u} = 0, \quad x^{(1)} = \frac{\partial H}{\partial \lambda_1}, \quad V_p^{(1)} = \frac{\partial H}{\partial \lambda_p}, \quad \lambda_1^{(1)} = -\frac{\partial H}{\partial x}, \quad \lambda_p^{(1)} = -\frac{\partial H}{\partial V_p},$$

that is, we need to solve the system of differential equations

$$\begin{cases} x^{(1)}(t) = \frac{B^{-1}}{\Gamma(\frac{7-t}{4})} - \frac{1}{2}B^{-2}t^{2\alpha(t)-2}\lambda_1(t) - AB^{-1}t^{-1}x(t) \\ \quad - \sum_{k=2}^{N} B^{-1}C_k t^{-k}V_k(t), \\ V_k^{(1)}(t) = (k-1)t^{k-2}x(t), \quad k = 2, \ldots, N, \\ \lambda_1^{(1)}(t) = AB^{-1}t^{-1}\lambda_1 - \sum_{k=2}^{N} (k-1)t^{k-2}\lambda_k(t), \\ \lambda_k^{(1)}(t) = B^{-1}C_k t^{-k}\lambda_1, \quad k = 2, \ldots, N, \end{cases} \quad (11.18)$$

subject to the boundary conditions

$$\begin{cases} x(0) = 0, \\ V_k(0) = 0, \quad k = 2, \ldots, N, \\ x(1) = 1, \\ \lambda_k(1) = 0, \quad k = 2, \ldots, N. \end{cases} \tag{11.19}$$

Figure 11.5 plots the numerical approximation $\tilde{x}_2(t)$ to the global minimizer $\overline{x}(t) = t$ of the variable order fractional problem of the calculus of variations (11.15) and (11.16), obtained solving (11.18) and (11.19) with $N = 2$. The approximation $\tilde{x}_2(t)$ is already visually indistinguishable from the exact solution $\overline{x}(t) = t$, and we do not increase the value of N. The effectiveness of our approach is also illustrated in Table 11.2, where some numerical values of the approximation $\tilde{x}_2(t)$ are given.

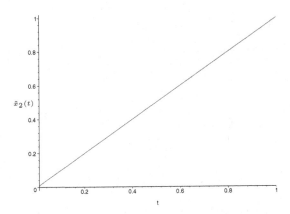

Fig. 11.5   Approximation $\tilde{x}_2(t)$ to the exact solution $\overline{x}(t) = t$ of the fractional problem of the calculus of variations (11.15) and (11.16), obtained from the application of Theorem 11.4 and the classical Pontryagin maximum principle, which is obtained solving (11.18) and (11.19) with $N = 2$.

Table 11.2   Some numerical values of the solution $\tilde{x}_2(t)$ of (11.18)–(11.19) with $N = 2$, close to the values of the global minimizer $\overline{x}(t) = t$ of the fractional variational problem of variable order (11.15) and (11.16).

t	0.2	0.4	0.6	0.8	1
$\tilde{x}_2(t)$	0.1998346692	0.3999020706	0.5999392936	0.7999708526	1.0000000000

Chapter 12

Discrete-time fractional calculus of variations

In this chapter we introduce a discrete-time fractional calculus of variations on the time scales $\mathbb{Z}$ and $(h\mathbb{Z})_a$. First and second-order necessary optimality conditions are established. Some numerical examples illustrating the use of the new Euler–Lagrange and Legendre type conditions are given.

12.1 Introduction

It is well known that discrete analogues of differential equations can be very useful in applications (Baleanu and Jarad, 2006; Jarad and Baleanu, 2007; Kelley and Peterson, 1991) and that fractional Euler–Lagrange differential equations are extremely difficult to solve, being necessary to discretize them (Agrawal, 2008b; Baleanu, Defterli and Agrawal, 2009). Here we consider, based on (Bastos, Ferreira and Torres, 2011a,b) (see also (Bastos, 2012)), a fractional discrete-time theory of the calculus of variations in a different time scale than $\mathbb{R}$. We dedicate two sections to that: one to the time scale $\mathbb{Z}$ and another to the time scale $(h\mathbb{Z})_a$. For an overview of fractional h-difference operators see (Mozyrska and Girejko, 2013); for applications of fractional difference calculus in control theory we refer the reader to (Mozyrska, 2014).

A time scale is any nonempty closed subset of the real line. The theory of time scales is a fairly new area of research. It was introduced in Stefan Hilger's 1988 Ph.D. thesis (Hilger, 1988) and subsequent landmark papers (Hilger, 1997, 1990), as a way to unify the seemingly disparate fields of discrete dynamical systems (i.e., difference equations) and continuous dynamical systems (i.e., differential equations). His dissertation referred to such unification as "Calculus on Measure Chains" (Aulbach and Hilger, 1990; Lakshmikantham, Sivasundaram and Kaymakcalan, 1996). Today it is better known as the time scale calculus. Since the nineties of the twentieth century, the study of dynamic equations on time scales received a lot of attention (see, e.g., (Agarwal *et al.*, 2002; Bohner and Peterson, 2001, 2003)). In 1997, the German mathematician Martin Bohner came across time scale calculus by chance, when he took up a position at the National University of Singapore. On the way from Singapore airport, a colleague, Ravi Agarwal, mentioned that time scale calculus might be the key to the problems that Bohner was investigating at that time. After that episode, time scale calculus became one of its main areas of research. To the reader who wants a gentle overview of time scales we advise to begin with (Spedding, 2003), which is written in a didactic way and at the same time points to some possible applications (e.g., in biology). Here we are interested in the calculus of variations on time scales (Dryl, Malinowska and Torres, 2013; Dryl and Torres, 2013; Martins and Torres, 2011; Torres, 2010) and to connect the theories of calculus of variations on time scales and fractional calculus.

12.2   Discrete fractional calculus

In contrast with the theory of continuous fractional calculus, which is well developed (Samko, Kilbas and Marichev, 1993), the theory of discrete fractional calculus is still in its infancy (Atici and Eloe, 2007, 2009; Miller and Ross, 1989). The fractional discrete theory has its foundations in the pioneering work of Kuttner in 1957, where the first definition of fractional-order differences appears. Here we review some definitions of fractional discrete operators.

In 1957 (Kuttner, 1957), Kuttner defined, for any sequence of complex numbers, $\{a_n\}$, the sth order difference as

$$\Delta^s a_n = \sum_{m=0}^{\infty} \binom{-s-1+m}{m} a_{n+m}, \tag{12.1}$$

where

$$\binom{t}{m} = \frac{t(t-1)\dots(t-m+1)}{m!}. \tag{12.2}$$

In (Kuttner, 1957) Kuttner also remarks that $\binom{t}{m}$ means 0 when m is negative and also when $t - m$ is a negative integer but t is not a negative integer. Clearly, (12.1) only makes sense when the series converges.

In 1974, Díaz and Osler (Díaz and Osler, 1974) gave the following definition for a fractional difference of order ν:

$$\Delta^\nu f(x) = \sum_{k=0}^\infty (-1)^k \binom{\nu}{k} f(x + \nu - k),$$

$$\binom{\nu}{k} = \frac{\Gamma(\nu + 1)}{\Gamma(\nu - k + 1)k!}, \tag{12.3}$$

where ν is any real or complex number. The above definition uses the usual well-known Gamma function (3.1). Throughout this chapter we use some of its most important properties:

$$\Gamma(1) = 1, \tag{12.4}$$

$$\Gamma(\nu + 1) = \nu\Gamma(\nu) \quad \text{for } \nu \in \mathbb{C}\backslash(-\mathbb{N}_0), \tag{12.5}$$

$$\Gamma(x + 1) = x!, \quad \text{for } x \in \mathbb{N}_0 . \tag{12.6}$$

Remark 12.1. Using the properties of Gamma function it is easily proved that formulas (12.2) and (12.3) coincide for ν an integer.

We begin by introducing some notation used throughout. Let a be an arbitrary real number and $b = a + k$ for a certain $k \in \mathbb{N}$ with $k \geq 2$. Let $\mathbb{T} = \{a, a + 1, \ldots, b\}$. In accordance with (Bohner and Peterson, 2001), we define the factorial function

$$t^{(n)} = t(t - 1)(t - 2)\ldots(t - n + 1), \quad n \in \mathbb{N}$$

and $t^{(0)} = 0$.

Extending the above definition from an integer n to an arbitrary real number α, we have

$$t^{(\alpha)} = \frac{\Gamma(t + 1)}{\Gamma(t + 1 - \alpha)}, \tag{12.7}$$

where Γ is the Euler Gamma function. Miller and Ross (1989) define a fractional sum of order $\nu > 0$ via the solution of a linear difference equation. Namely, they present it as follows:

Definition 12.1.

$$\Delta^{-\nu} f(t) = \frac{1}{\Gamma(\nu)} \sum_{s=a}^{t-\nu} (t - \sigma(s))^{(\nu-1)} f(s). \tag{12.8}$$

Here f is defined for $s = a \mod (1)$ and $\Delta^{-\nu} f$ is defined for $t = (a + \nu) \mod (1)$.

This was done in analogy with the Riemann–Liouville fractional integral of order $\nu > 0$, which can be obtained via the solution of a linear differential equation (Miller and Ross, 1989, 1993). Some basic properties of the sum in (12.8) were obtained in (Miller and Ross, 1989). Although there are other definitions of fractional difference operators throughout this chapter, we follow mostly the spirit of Miller and Ross, Atici and Eloe (Atici and Eloe, 2009; Miller and Ross, 1989).

12.3　Basic definitions on time scales

Here we recall some basic results on time scales that are used in the sequel. In Section 12.4 we review some results of the calculus of variations on time scales.

Definition 12.2. *A time scale is an arbitrary nonempty closed subset of* $\mathbb{R}$ *and is denoted by* $\mathbb{T}$.

Example 12.1. Here we just give some examples of sets that are time scales and others that are not.

- The set $\mathbb{R}$ is a time scale;
- The set $\mathbb{Z}$ is a time scale;
- The Cantor set is a time scale;
- The set $\mathbb{C}$ is not a time scale;
- The set $\mathbb{Q}$ is not a time scale.

In applications, the quantum calculus that has, as base, the set of powers of a given number q is, in addition to classical continuous and purely discrete calculus, one of the most important time scales. Applications of this calculus appear, for example, in physics (Kac and Cheung, 2002) and economics (Malinowska and Torres, 2010c, 2014). The quantum derivative was introduced by Leonhard Euler and a fractional formulation of that derivative can be found in (Ortigueira, 2010a,b).

In this chapter attention is given to time scales $\mathbb{Z}$, $\mathbb{R}$ and $(h\mathbb{Z})_a = \{a, a+h, a+2h, \ldots\}$, $a \in \mathbb{R}$, $h > 0$.

A time scale of the form of a union of disjoint closed real intervals, constitutes a good background for the study of population models (of plants, insects, etc.). Such models appear, for example, when a plant population exhibits exponential growth during the months of Spring and Summer and, at the beginning of Autumn, when plants die, the seeds remain in the ground. Similar examples concerning insect populations, where all the adults die before the babies are born, can be found in (Bohner and Peterson, 2001; Spedding, 2003).

The following operators of time scales theory are used several times.

Definition 12.3. *The mapping* σ : $\mathbb{T} \longrightarrow \mathbb{T}$, *defined by* $\sigma(t) =$ $\inf \{s \in \mathbb{T} : s > t\}$ *with* $\inf \emptyset = \sup \mathbb{T}$ *(i.e., $\sigma(M) = M$ if $\mathbb{T}$ has a maximum M) is called the* forward jump operator. *Accordingly, we define the backward jump operator* ρ : $\mathbb{T} \to \mathbb{T}$ *by* $\rho(t) = \sup \{s \in \mathbb{T} : s < t\}$ *with* $\sup \emptyset = \inf \mathbb{T}$ *(i.e., $\rho(m) = m$ if $\mathbb{T}$ has a minimum m). The symbol $\emptyset$ denotes the empty set.*

The following classification of points is used within the theory: a point $t \in \mathbb{T}$ is called right-dense, right-scattered, left-dense or left-scattered if $\sigma(t) = t$, $\sigma(t) > t$, $\rho(t) = t$, $\rho(t) < t$, respectively. A point t is called isolated if $\rho(t) < t < \sigma(t)$ and dense if $\rho(t) = t = \sigma(t)$.

Definition 12.4. *A function $f : \mathbb{T} \to \mathbb{R}$ is called regulated provided its right-sided limits exist (finite) at all right-dense points in $\mathbb{T}$ and its left-sided limits exist (finite) at all left-dense points in $\mathbb{T}$.*

Example 12.2. Let $\mathbb{T} = \{0\} \cup \left\{ \dfrac{1}{n} : n \in \mathbb{N} \right\} \cup \{2\} \cup \left\{ 2 - \dfrac{1}{n} : n \in \mathbb{N} \right\}$ and define $f : \mathbb{T} \to \mathbb{R}$ by

$$f(t) := \begin{cases} 0 \text{ if } t \neq 2; \\ t \text{ if } t = 2. \end{cases}$$

Function f is regulated.

Now, let us define the sets $\mathbb{T}^{\kappa^n}$, inductively:

$$\mathbb{T}^{\kappa^1} = \mathbb{T}^{\kappa} = \{t \in \mathbb{T} : t \text{ non-maximal or left-dense}\}$$

and $\mathbb{T}^{\kappa^n} = (\mathbb{T}^{\kappa^{n-1}})^{\kappa}$, $n \geq 2$.

Definition 12.5. *The forward graininess function $\mu : \mathbb{T} \to [0, \infty)$ is defined by $\mu(t) = \sigma(t) - t$.*

Remark 12.2. Throughout the chapter, we refer to the forward graininess function simply as the graininess function.

Definition 12.6. *The backward graininess function $\nu : \mathbb{T} \to [0, \infty)$ is defined by $\nu(t) = t - \rho(t)$.*

Example 12.3. If $\mathbb{T} = \mathbb{R}$, then $\sigma(t) = \rho(t) = t$ and $\mu(t) = \nu(t) = 0$. If $\mathbb{T} = [-4, 1] \cup \mathbb{N}$, then

$$\sigma(t) = \begin{cases} t & \text{if } t \in [-4, 1); \\ t + 1 & \text{otherwise,} \end{cases}$$

while

$$\rho(t) = \begin{cases} t & \text{if } t \in [-4, 1]; \\ t - 1 & \text{otherwise.} \end{cases}$$

Moreover,

$$\mu(t) = \begin{cases} 0 & \text{if } t \in [-4, 1); \\ 1 & \text{otherwise,} \end{cases}$$

and

$$\nu(t) = \begin{cases} 0 & \text{if } t \in [-4, 1]; \\ 1 & \text{otherwise.} \end{cases}$$

Example 12.4. If $\mathbb{T} = \mathbb{Z}$, then $\sigma(t) = t+1$, $\rho(t) = t-1$ and $\mu(t) = \nu(t) = 1$. If $\mathbb{T} = (h\mathbb{Z})_a$ then $\sigma(t) = t + h$, $\rho(t) = t - h$ and $\mu(t) = \nu(t) = h$.

For two points $a, b \in \mathbb{T}$, the time scale interval is defined by

$$[a, b]_{\mathbb{T}} = \{t \in \mathbb{T} : a \leq t \leq b\}.$$

Throughout the chapter we will frequently write $f^\sigma(t) = f(\sigma(t))$ and $f^\rho(t) = f(\rho(t))$.

Next results are related with differentiation on time scales.

Definition 12.7. *We say that a function $f : \mathbb{T} \to \mathbb{R}$ is Δ-differentiable at $t \in \mathbb{T}^\kappa$ if there is a number $f^\Delta(t)$ such that for all $\varepsilon > 0$ there exists a neighborhood U of t such that*

$$|f^\sigma(t) - f(s) - f^\Delta(t)(\sigma(t) - s)| \leq \varepsilon|\sigma(t) - s| \quad \text{for all } s \in U.$$

We call $f^\Delta(t)$ the Δ-derivative of f at t.

The Δ-derivative of order $n \in \mathbb{N}$ of a function f is defined by $f^{\Delta^n}(t) = \left(f^{\Delta^{n-1}}(t) \right)^\Delta$, $t \in \mathbb{T}^{\kappa^n}$, provided the right-hand side of the equality exists, where $f^{\Delta^0} = f$.

Some basic properties will now be given for the Δ-derivative.

Theorem 12.1 (Theorem 1.16 of (Bohner and Peterson, 2001)). *Assume $f : \mathbb{T} \to \mathbb{R}$ is a function and let $t \in \mathbb{T}^\kappa$. Then we have the following:*

(1) If f is Δ-differentiable at t, then f is continuous at t.

(2) If f is continuous at t and t is right-scattered, then f is Δ-differentiable at t with

$$f^\Delta(t) = \frac{f^\sigma(t) - f(t)}{\mu(t)}. \tag{12.9}$$

(3) If t is right-dense, then f is Δ-differentiable at t if and only if the limit

$$\lim_{s \to t} \frac{f(s) - f(t)}{s - t}$$

exists as a finite number. In this case,

$$f^\Delta(t) = \lim_{s \to t} \frac{f(s) - f(t)}{s - t}.$$

(4) If f is Δ-differentiable at t, then

$$f^\sigma(t) = f(t) + \mu(t) f^\Delta(t). \tag{12.10}$$

It is an immediate consequence of Theorem 12.1 that if $\mathbb{T} = \mathbb{R}$, then the Δ-derivative becomes the classical one, i.e., $f^\Delta = f'$ while if $\mathbb{T} = \mathbb{Z}$, the Δ-derivative reduces to the forward difference $f^\Delta(t) = \Delta f(t) = f(t+1) - f(t)$.

Theorem 12.2 (Theorem 1.20 of (Bohner and Peterson, 2001)). *Assume $f, g : \mathbb{T} \to \mathbb{R}$ are Δ-differentiable at $t \in \mathbb{T}^\kappa$. Then:*

(1) The sum $f + g : \mathbb{T} \to \mathbb{R}$ is Δ-differentiable at t and $(f + g)^\Delta(t) = f^\Delta(t) + g^\Delta(t)$.

(2) For any number $\xi \in \mathbb{R}$, $\xi f : \mathbb{T} \to \mathbb{R}$ is Δ-differentiable at t and $(\xi f)^\Delta(t) = \xi f^\Delta(t)$.

(3) The product $fg : \mathbb{T} \to \mathbb{R}$ is Δ-differentiable at t and

$$(fg)^\Delta(t) = f^\Delta(t)g(t) + f^\sigma(t)g^\Delta(t) = f(t)g^\Delta(t) + f^\Delta(t)g^\sigma(t). \tag{12.11}$$

Definition 12.8. *A function $f : \mathbb{T} \to \mathbb{R}$ is called rd-continuous if it is continuous at right-dense points and if the left-sided limit exists at left-dense points.*

Remark 12.3. We denote the set of all rd-continuous functions by C_{rd} or $C_{rd}(\mathbb{T})$, and the set of all Δ-differentiable functions with rd-continuous derivative by C_{rd}^1 or $C_{rd}^1(\mathbb{T})$.

Example 12.5. Let $\mathbb{T} = \mathbb{N}_0 \cup \left\{ 1 - \frac{1}{n} : n \in \mathbb{N} \right\}$ and define $f : \mathbb{T} \to \mathbb{R}$ by

$$f(t) := \begin{cases} 0 \text{ if } t \in \mathbb{N}; \\ t \text{ otherwise.} \end{cases}$$

It is easy to verify that f is continuous at the isolated points. So, the points that need our careful attention are the right-scattered point 0 and the left-dense point 1. The right-sided limit of f at 0 exist and is finite (is equal to 0). The left-sided limit of f at 1 exist and is finite (is equal to 1). We conclude that f is rd-continuous in $\mathbb{T}$.

We consider now some results about integration on time scales.

Definition 12.9. *A function $F : \mathbb{T} \to \mathbb{R}$ is called an antiderivative of $f : \mathbb{T}^\kappa \to \mathbb{R}$ provided $F^\Delta(t) = f(t)$ for all $t \in \mathbb{T}^\kappa$.*

Theorem 12.3 (Theorem 1.74 of (Bohner and Peterson, 2001)).
Every rd-continuous function has an antiderivative.

Let $f : \mathbb{T}^\kappa :\to \mathbb{R}$ be a rd-continuous function and let $F : \mathbb{T} \to \mathbb{R}$ be an antiderivative of f. Then, the Δ-integral is defined by $\int_a^b f(t)\Delta t = F(b) - F(a)$ for all $a, b \in \mathbb{T}$.

One can easily prove (Bohner and Peterson, 2001, Theorem 1.79) that, when $\mathbb{T} = \mathbb{R}$ then $\int_a^b f(t)\Delta t = \int_a^b f(t)dt$, being the right-hand side of the equality the usual Riemann integral, and when $[a, b] \cap \mathbb{T}$ contains only isolated points, then

$$\int_a^b f(t)\Delta t = \begin{cases} \sum_{t=a}^{\rho(b)} \mu(t)f(t) & \text{if } a < b; \\ 0 & \text{if } a = b; \\ -\sum_{t=a}^{\rho(b)} \mu(t)f(t), & \text{if } a > b. \end{cases} \quad (12.12)$$

Remark 12.4. When $\mathbb{T} = \mathbb{Z}$ or $\mathbb{T} = h\mathbb{Z}$, equation (12.12) holds.

The $\Delta - integral$ also satisfies

$$\int_t^{\sigma(t)} f(\tau)\Delta\tau = \mu(t)f(t), \quad t \in \mathbb{T}^\kappa. \quad (12.13)$$

Theorem 12.4 (Theorem 1.77 of (Bohner and Peterson, 2001)).
Let $a, b, c \in \mathbb{T}$, $\xi \in \mathbb{R}$ and $f, g \in C_{rd}(\mathbb{T}^\kappa)$. Then,

(1) $\int_a^b [f(t) + g(t)]\Delta t = \int_a^b f(t)\Delta t + \int_a^b g(t)\Delta t$.
(2) $\int_a^b (\xi f)(t)\Delta t = \xi \int_a^b f(t)\Delta t$.
(3) $\int_a^b f(t)\Delta t = -\int_b^a f(t)\Delta t$.
(4) $\int_a^b f(t)\Delta t = \int_a^c f(t)\Delta t + \int_c^b f(t)\Delta t$.

(5) $\int_a^a f(t)\Delta t = 0$.

(6) If $|f(t)| \le g(t)$ on $[a,b]_{\mathbb{T}}^\kappa$, then

$$\left| \int_a^b f(t)\Delta t \right| \le \int_a^b g(t)\Delta t.$$

We now present the integration by parts formulas for the Δ-integral:

Lemma 12.1 (cf. Theorem 1.77 of (Bohner and Peterson, 2001)).
If $a,b \in \mathbb{T}$ and $f,g \in C_{rd}^1$, then

$$\int_a^b f^\sigma(t)g^\Delta(t)\Delta t = [(fg)(t)]_{t=a}^{t=b} - \int_a^b f^\Delta(t)g(t)\Delta t; \qquad (12.14)$$

$$\int_a^b f(t)g^\Delta(t)\Delta t = [(fg)(t)]_{t=a}^{t=b} - \int_a^b f^\Delta(t)g^\sigma(t)\Delta t. \qquad (12.15)$$

Remark 12.5. For analogous results on ∇-integrals the reader can consult, e.g., (Bohner and Peterson, 2003).

Some more definitions and results must be presented since we will need them. We start defining functions that generalize polynomials functions in $\mathbb{R}$ to the times scales calculus. There are, at least, two ways of doing that:

Definition 12.10. *We define the polynomials on time scales, $g_k, h_k : \mathbb{T}^2 \to \mathbb{R}$ for $k \in \mathbb{N}_0$ as follows:*

$$g_0(t,s) = h_0(t,s) \equiv 1 \quad \forall s,t \in \mathbb{T}$$

$$g_{k+1}(t,s) = \int_s^t g_k(\sigma(\tau),s)\Delta\tau \quad \forall s,t \in \mathbb{T},$$

$$h_{k+1}(t,s) = \int_s^t h_k(\tau,s)\Delta\tau \quad \forall s,t \in \mathbb{T}.$$

Remark 12.6. If we let $h_k^\Delta(t,s)$ denote, for each fixed $s \in \mathbb{T}$, the derivative of $h_k(t,s)$ with respect to t, then we have the following equalities:

(1) $h_k^\Delta(t,s) = h_{k-1}(t,s)$ for $k \in \mathbb{N}$, $t \in \mathbb{T}^\kappa$;

(2) $g_k^\Delta(t,s) = g_{k-1}(\sigma(t),s)$ for $k \in \mathbb{N}$, $t \in \mathbb{T}^\kappa$;

(3) $g_1(t,s) = h_1(t,s) = t - s$ for all $s,t \in \mathbb{T}$.

Remark 12.7. It is not easy to find the explicit form of g_k and h_k for a generic time scale.

To give the reader an idea what is the formula of such polynomials for particular time scales, we give the following two examples:

Example 12.6. For $\mathbb{T} = \mathbb{R}$ and $k \in \mathbb{N}_0$ we have:

$$g_k(t,s) = h_k(t,s) = \frac{(t-s)^k}{k!} \quad \text{for all} \quad s,t \in \mathbb{T}. \qquad (12.16)$$

Example 12.7. Consider $\mathbb{T} = \overline{q^{\mathbb{Z}}} = \{q^k : k \in \mathbb{Z}\} \cup \{0\}$. For $k \in \mathbb{N}_0$ and $q > 1$ we have:

$$h_k(t,s) = \prod_{i=0}^{k-1} \frac{t - q^i s}{\sum_{j=0}^i q^j} \quad \text{for all} \quad s,t \in \mathbb{T}. \qquad (12.17)$$

12.4 Calculus of variations on time scales

In (Bohner, 2004), Martin Bohner initiated the theory of calculus of variations on time scales. This section is dedicated to presenting some results of this theory.

Definition 12.11. *A function f defined on $[a,b]_{\mathbb{T}} \times \mathbb{R}$ is called continuous in the second variable, uniformly in the first variable, if for each $\varepsilon > 0$ there exists $\delta > 0$ such that $|x_1 - x_2| < \delta$ implies $|f(t,x_1) - f(t,x_2)| < \varepsilon$ for all $t \in [a,b]_{\mathbb{T}}$.*

Lemma 12.2 (cf. Lemma 2.2 of (Bohner, 2004)).
Suppose that $F[x] = \int_a^b f(t,x(t))\Delta t$ is well defined. If f_x is continuous in x, uniformly in t, then $F'[x] = \int_a^b f_x(t,x(t))\Delta t$.

Let $a,b \in \mathbb{T}$ and $L(t,u,v) : [a,b]_{\mathbb{T}}^{\kappa} \times \mathbb{R}^2 \to \mathbb{R}$. *Consider the problem of finding a function $y \in C_{rd}^1$ such that*

$$\mathcal{L}[y] = \int_a^b L(t, y^{\sigma}(t), y^{\Delta}(t))\Delta t \longrightarrow \min, \quad y(a) = y_a, \quad y(b) = y_b, \quad (12.18)$$

with $y_a, y_b \in \mathbb{R}$.

Remark 12.8. If we fix $\mathbb{T} = \mathbb{R}$ problem (12.18) reduces to the classical basic problem of the calculus of variations (van Brunt, 2004).

Remark 12.9. If we fix $\mathbb{T} = \mathbb{Z}$ we obtain the discrete version of (12.18). The goal is to find a function $y \in C^1_{rd}$ such that

$$\mathcal{L}[y] = \sum_{t=a}^{b-1} L(t, y(t+1), y^{\Delta}(t))\Delta t \longrightarrow \min, \quad y(a) = y_a, \quad y(b) = y_b,$$

$$(12.19)$$

where a, $b \in \mathbb{Z}$ with $a < b$; $y_a, y_b \in \mathbb{R}$ and $L : \mathbb{Z} \times \mathbb{R}^2 \to \mathbb{R}$.

Definition 12.12. *For $f \in C^1_{rd}$ we define the norm*

$$\|f\| = \sup_{t \in [a,b]^{\kappa}_{\mathbb{T}}} |f^{\sigma}(t)| + \sup_{t \in [a,b]^{\kappa}_{\mathbb{T}}} |f^{\Delta}(t)|.$$

A function $\hat{y} \in C^1_{rd}$ with $\hat{y}(a) = y_a$ and $\hat{y}(b) = y_b$ is called a (weak) local minimizer for problem (12.18) provided there exists $\delta > 0$ such that $\mathcal{L}[\hat{y}] \leq \mathcal{L}[y]$ for all $y \in C^1_{rd}$ with $y(a) = y_a$ and $y(b) = y_b$ and $\|y - \hat{y}\| < \delta$.

Definition 12.13. *A function $\eta \in C^1_{rd}$ is called an admissible variation provided $\eta \neq 0$ and $\eta(a) = \eta(b) = 0$.*

Lemma 12.3 (cf. Lemma 3.4 of (Bohner, 2004)). *Let $y, \eta \in C^1_{rd}$ be arbitrary fixed functions. Put $f(t, \varepsilon) = L(t, y^{\sigma}(t) + \varepsilon\eta^{\sigma}(t), y^{\Delta}(t) + \varepsilon\eta^{\Delta}(t))$ and $\Phi(\varepsilon) = \mathcal{L}[y + \varepsilon\eta]$, $\varepsilon \in \mathbb{R}$. If f_{ε} and $f_{\varepsilon\varepsilon}$ are continuous in ε, uniformly in t, then*

$$\Phi'(\varepsilon) = \int_a^b [L_u(t, y^{\sigma}(t), y^{\Delta}(t))\eta^{\sigma}(t) + L_v(t, y^{\sigma}(t), y^{\Delta}(t))\eta^{\Delta}(t)]\Delta t,$$

$$\Phi''(\varepsilon) = \int_a^b \{L_{uu}[y](t)(\eta^{\sigma}(t))^2 + 2\eta^{\sigma}(t)L_{uv}[y](t)\eta^{\Delta}(t) + L_{vv}[y](t)(\eta^{\Delta}(t))^2\}\Delta t,$$

where $[y](t) = (t, y^{\sigma}(t), y^{\Delta}(t))$.

The next lemma is the time scales version of the classical Dubois–Reymond lemma.

Lemma 12.4 (cf. Lemma 4.1 of (Bohner, 2004)).
Let $g \in C_{rd}([a,b]^{\kappa}_{\mathbb{T}})$. Then,

$$\int_a^b g(t)\eta^{\Delta}(t)\Delta t = 0, \quad \text{for all } \eta \in C^1_{rd}([a,b]_{\mathbb{T}}) \text{ with } \eta(a) = \eta(b) = 0,$$

holds if and only if

$$g(t) = c, \quad \text{on } [a,b]^{\kappa}_{\mathbb{T}} \text{ for some } c \in \mathbb{R}.$$

The following theorem give first and second-order necessary optimality conditions for problem (12.18).

Theorem 12.5 (cf. (Bohner, 2004)). *Suppose that L satisfies the assumption of Lemma 12.3. If $\hat{y} \in C_{rd}^1$ is a (weak) local minimizer for problem given by (12.18), then necessarily*

(1) $L_v^\Delta[\hat{y}](t) = L_u[\hat{y}](t)$, $t \in [a,b]_{\mathbb{T}}^\kappa$ *(time scale Euler–Lagrange equation).*

(2) $L_{vv}[\hat{y}](t) + \mu(t)\{2L_{uv}[\hat{y}](t) + \mu(t)L_{uu}[\hat{y}](t) + (\mu^\sigma(t))^* L_{vv}[\hat{y}](\sigma(t))\} \geq 0$,
$t \in [a,b]_{\mathbb{T}}^{\kappa^2}$ *(time scale Legendre's condition),*

where $[y](t) = (t, y^\sigma(t), y^\Delta(t))$ *and* $\alpha^* = \frac{1}{\alpha}$ *if* $\alpha \in \mathbb{R}\backslash\{0\}$ *and* $0^* = 0$.

Example 12.8 (Example 4.3 of (Bohner, 2004)). Find the solution of the problem

$$\int_a^b \sqrt{1 + (y^\Delta)^2}\Delta t \to \min, \qquad y(a) = 0, \quad y(b) = 1. \tag{12.20}$$

Writing (12.20) in form of (12.18) we have

$$L(t,u,v) = \sqrt{1 + v^2}, \qquad L_u(t,u,v) = 0 \quad \text{and} \quad L_v(t,u,v) = \frac{v}{1+v^2}.$$

Suppose $\hat{y}$ is a local minimizer of (12.20). Then, by item 1 of Theorem 12.5, equation

$$L_v^\Delta(t, \hat{y}^\sigma(t), \hat{y}^\Delta(t)) = 0, \quad t \in [a,b]^\kappa, \tag{12.21}$$

must hold. The last equation implies that there exists a constant $c \in \mathbb{R}$ such that

$$L_v(t, \hat{y}^\sigma(t), \hat{y}^\Delta(t)) \equiv c, \quad t \in [a,b],$$

i.e.,

$$\hat{y}^\Delta(t) = c\sqrt{1 + (\hat{y}^\Delta)^2}, \quad t \in [a,b] \tag{12.22}$$

holds. Solving equation (12.22) we obtain

$$\hat{y}(t) = \frac{t - a}{b - a} \quad \text{for all} \quad t \in [a,b],$$

which is the equation of the straight line connecting the given points.

We refer the reader to (Bartosiewicz and Torres, 2008; Bastos, Mozyrska and Torres, 2011; Bastos and Torres, 2010; Bohner, Ferreira and Torres, 2010; Ferreira, 2010; Hilscher and Zeidan, 2004; Malinowska and Torres, 2010b) and references therein for results on calculus of variations on time scales. Here our main interest on the subject is just as the starting point to the fractional case on the time scale $\mathbb{Z}$ (Section 12.5) and $\mathbb{T} = (h\mathbb{Z})_a$ (Section 12.6).

12.5 Fractional variational problems in $\mathbb{T} = \mathbb{Z}$

In this section we introduce a discrete-time fractional calculus of variations on the time scale $\mathbb{Z}$. First and second-order necessary optimality conditions are established. We finish the chapter with some examples illustrating the use of the new Euler–Lagrange and Legendre type conditions.

12.5.1 *Introduction*

As mentioned in Section 12.2, Miller and Ross define the fractional sum of order $\nu > 0$ by

$$\Delta^{-\nu} f(t) = \frac{1}{\Gamma(\nu)} \sum_{s=a}^{t-\nu} (t - \sigma(s))^{(\nu-1)} f(s). \tag{12.23}$$

Remark 12.10. This was done in analogy with the left Riemann–Liouville fractional integral of order $\nu > 0$,

$$_a I_x^{-\nu} f(x) = \frac{1}{\Gamma(\nu)} \int_a^x (x - s)^{\nu-1} f(s) ds,$$

which can be obtained via the solution of a linear differential equation (Miller and Ross, 1989, 1993). Some basic properties of the sum in (12.23) were obtained in (Miller and Ross, 1989).

More recently, Atici and Eloe (2007, 2009) defined the fractional difference of order $\alpha > 0$, i.e., $\Delta^\alpha f(t) = \Delta^m (\Delta^{-(m-\alpha)} f(t))$ with m the least integer satisfying $m \geq \alpha$, and developed some of its properties to allow solutions of certain fractional difference equations.

To the best of the author's knowledge, the first paper with a theory for the discrete calculus of variations (using forward difference operator) was written by Tomlinson Fort in 1937 (Fort, 1937). Some important results of discrete calculus of variations are summarized in (Kelley and Peterson, 1991, Chap. 8). It is well known that discrete analogues of differential equations can be very useful in applications (Hilscher and Zeidan, 2005; Kelley and Peterson, 1991). Therefore, we consider pertinent to start here a fractional discrete-time theory of the calculus of variations.

Throughout the chapter we proceed to develop the theory of fractional difference calculus, namely, we introduce the concept of left and right fractional sum/difference (cf. Definitions 12.14 and 12.15 below) and prove some new results related to them.

We begin, in Section 12.5.2, to give the definitions and results needed throughout. In Section 12.5.3 we present and prove the new results; in Section 12.5.5 we give some examples. Finally, in Section 12.5.6 we mention the main conclusions of the chapter, and some possible extensions and open questions. Computer code done in the Computer Algebra System Maxima is available in (Bastos, 2012).

12.5.2   *Preliminaries*

We begin by introducing some notation used throughout. Let a be an arbitrary real number and $b = k + a$ for a certain $k \in \mathbb{N}$ with $k \geq 2$. In this chapter we restrict ourselves to the time scale $\mathbb{T} = \{a, a+1, \ldots, b\}$.

Denote by $\mathcal{F}$ the set of all real valued functions defined on $\mathbb{T}$. According to Example 12.4, we have $\sigma(t) = t + 1$, $\rho(t) = t - 1$.

The usual conventions $\sum_{t=c}^{c-1} f(t) = 0$, $c \in \mathbb{T}$, and $\prod_{i=0}^{-1} f(i) = 1$ remain valid here. As usual, the forward difference is defined by $\Delta f(t) = f^{\sigma}(t) - f(t)$. If we have a function f of two variables, $f(t,s)$, its partial (difference) derivatives are denoted by Δ_t and Δ_s, respectively. Recalling (12.7) we can, for arbitrary $x, y \in \mathbb{R}$, define (when it makes sense)

$$x^{(y)} = \frac{\Gamma(x+1)}{\Gamma(x+1-y)} \tag{12.24}$$

where Γ is the Gamma function.

While reaching the proof of Theorem 12.8 we actually "find" the definition of the left and right fractional sum:

Definition 12.14. *Let $f \in \mathcal{F}$. The left fractional sum and the right fractional sum of order $\nu > 0$ are defined, respectively, as*

$$_a\Delta_t^{-\nu} f(t) = \frac{1}{\Gamma(\nu)} \sum_{s=a}^{t-\nu} (t - \sigma(s))^{(\nu-1)} f(s), \tag{12.25}$$

and

$$_t\Delta_b^{-\nu} f(t) = \frac{1}{\Gamma(\nu)} \sum_{s=t+\nu}^{b} (s - \sigma(t))^{(\nu-1)} f(s). \tag{12.26}$$

Remark 12.11. The above sums (12.25) and (12.26) are defined for $t \in \{a+\nu, a+\nu+1, \ldots, b+\nu\}$ and $t \in \{a-\nu, a-\nu+1, \ldots, b-\nu\}$, respectively, while $f(t)$ is defined for $t \in \{a, a+1, \ldots, b\}$. Throughout we will write (12.25) and (12.26), respectively, in the following way:

$$_a\Delta_t^{-\nu} f(t) = \frac{1}{\Gamma(\nu)} \sum_{s=a}^{t} (t + \nu - \sigma(s))^{(\nu-1)} f(s), \quad t \in \mathbb{T},$$

$$_t\Delta_b^{-\nu} f(t) = \frac{1}{\Gamma(\nu)} \sum_{s=t}^{b} (s + \nu - \sigma(t))^{(\nu-1)} f(s), \quad t \in \mathbb{T}.$$

Remark 12.12. The left fractional sum defined in (12.25) coincides with the fractional sum defined in (Miller and Ross, 1989) (see also (12.8)). The analogy of (12.25) and (12.26) with the Riemann–Liouville left and right fractional integrals of order $\nu > 0$ is clear:

$$_aI_x^{-\nu} f(x) = \frac{1}{\Gamma(\nu)} \int_a^x (x - s)^{\nu-1} f(s) ds,$$

$$_xI_b^{-\nu} f(x) = \frac{1}{\Gamma(\nu)} \int_x^b (s - x)^{\nu-1} f(s) ds.$$

It was proved in (Miller and Ross, 1989) that $\lim_{\nu \to 0} {}_a\Delta_t^{-\nu} f(t) = f(t)$. We do the same for the right fractional sum using a different method. Let $\nu > 0$ be arbitrary. Then,

$$_t\Delta_b^{-\nu} f(t) = \frac{1}{\Gamma(\nu)} \sum_{s=t}^{b} (s + \nu - \sigma(t))^{(\nu-1)} f(s)$$

$$= f(t) + \frac{1}{\Gamma(\nu)} \sum_{s=\sigma(t)}^{b} (s + \nu - \sigma(t))^{(\nu-1)} f(s)$$

$$= f(t) + \sum_{s=\sigma(t)}^{b} \frac{\Gamma(s + \nu - t)}{\Gamma(\nu)\Gamma(s - t + 1)} f(s)$$

$$= f(t) + \sum_{s=\sigma(t)}^{b} \frac{\prod_{i=0}^{s-t-1} (\nu + i)}{\Gamma(s - t + 1)} f(s).$$

Therefore, $\lim_{\nu \to 0} {}_t\Delta_b^{-\nu} f(t) = f(t)$. It is now natural to define

$$_a\Delta_t^0 f(t) = {}_t\Delta_b^0 f(t) = f(t), \qquad (12.27)$$

which we do here, and to write

$$_a\Delta_t^{-\nu} f(t) = f(t) + \frac{\nu}{\Gamma(\nu+1)} \sum_{s=a}^{t-1} (t + \nu - \sigma(s))^{(\nu-1)} f(s), \quad t \in \mathbb{T}, \quad \nu \geq 0,$$

$$\qquad (12.28)$$

$$_t\Delta_b^{-\nu} f(t) = f(t) + \frac{\nu}{\Gamma(\nu+1)} \sum_{s=\sigma(t)}^{b} (s + \nu - \sigma(t))^{(\nu-1)} f(s), \quad t \in \mathbb{T}, \quad \nu \geq 0.$$

The next theorem was proved in (Atici and Eloe, 2009).

Theorem 12.6 (See (Atici and Eloe, 2009)). *Let $f \in \mathcal{F}$ and $\nu > 0$. Then, the equality*

$$_a\Delta_t^{-\nu}\Delta f(t) = \Delta(_a\Delta_t^{-\nu}f(t)) - \frac{(t+\nu-a)^{(\nu-1)}}{\Gamma(\nu)}f(a), \quad t \in \mathbb{T}^\kappa,$$

holds.

Remark 12.13. It is easy to include the case $\nu = 0$ in Theorem 12.6. Indeed, in view of (12.5) and (12.27), we get

$$_a\Delta_t^{-\nu}\Delta f(t) = \Delta(_a\Delta_t^{-\nu}f(t)) - \frac{\nu}{\Gamma(\nu+1)}(t+\nu-a)^{(\nu-1)}f(a), \quad t \in \mathbb{T}^\kappa,$$
(12.29)

for all $\nu \geq 0$.

Now, we prove the counterpart of Theorem 12.6 for the right fractional sum. The result was first given in (Bastos, Ferreira and Torres, 2011a, Theorem 2.3).

Theorem 12.7. *Let $f \in \mathcal{F}$ and $\nu \geq 0$. Then, the equality*

$$_t\Delta_{\rho(b)}^{-\nu}\Delta f(t) = \frac{\nu}{\Gamma(\nu+1)}(b+\nu-\sigma(t))^{(\nu-1)}f(b) + \Delta(_t\Delta_b^{-\nu}f(t)), \quad t \in \mathbb{T}^\kappa,$$
(12.30)

holds.

Proof. We only prove the case $\nu > 0$ as the case $\nu = 0$ is trivial (see Remark 12.13). We start by fixing an arbitrary $t \in \mathbb{T}^\kappa$ and prove that for all $s \in \mathbb{T}^\kappa$,

$$\Delta_s\left((s+\nu-\sigma(t))^{(\nu-1)}f(s)\right)$$
$$= (\nu-1)(s+\nu-\sigma(t))^{(\nu-2)}f^\sigma(s) + (s+\nu-\sigma(t))^{(\nu-1)}\Delta f(s). \quad (12.31)$$

By definition of forward difference with respect to variable s we can write that

$$\Delta_s\left((s+\nu-\sigma(t))^{(\nu-1)}f(s)\right)$$
$$= (\sigma(s)+\nu-\sigma(t))^{(\nu-1)}f^\sigma(s) - (s+\nu-\sigma(t))^{(\nu-1)}f(s)$$
$$= (s+\nu-t)^{(\nu-1)}f^\sigma(s) - (s+\nu-t-1)^{(\nu-1)}f(s) \quad (12.32)$$
$$= \frac{\Gamma(s+\nu-t+1)f^\sigma(s)}{\Gamma(s-t+2)} - \frac{\Gamma(s+\nu-t)f(s)}{\Gamma(s-t+1)}.$$

Applying (12.5) to the numerator and denominator of first fraction we have that (12.32) is equal to

$$
\frac{(s+\nu-t)\Gamma(s+\nu-t)f^\sigma(s)}{(s-t+1)\Gamma(s-t+1)} - \frac{\Gamma(s+\nu-t)f(s)}{\Gamma(s-t+1)}
$$
$$
= \frac{[(s+\nu-t)f^\sigma(s) - (s-t+1)f(s)]\,\Gamma(s+\nu-t)}{(s-t+1)\Gamma(s-t+1)}
$$
$$
= \frac{[(s-t+1)f^\sigma(s) + (\nu-1)f^\sigma(s) - (s-t+1)f(s)]\,\Gamma(s+\nu-t)}{(s-t+1)\Gamma(s-t+1)}
$$
$$
= (\nu-1)\frac{\Gamma(s+\nu-\sigma(t)+1)}{\Gamma(s+\nu-\sigma(t)-(\nu-2)+1)}f^\sigma(s)
$$
$$
\quad + \frac{\Gamma(s+\nu-\sigma(t)+1)}{\Gamma(s+\nu-\sigma(t)-(\nu-1)+1)}(f^\sigma(s) - f(s))
$$
$$
= (\nu-1)(s+\nu-\sigma(t))^{(\nu-2)}f^\sigma(s) + (s+\nu-\sigma(t))^{(\nu-1)}\Delta f(s).
$$

Now, with (12.31) proven, we can state that

$$
\frac{1}{\Gamma(\nu)}\sum_{s=t}^{b-1}(s+\nu-\sigma(t))^{(\nu-1)}\Delta f(s)
$$
$$
= \left[\frac{(s+\nu-\sigma(t))^{(\nu-1)}}{\Gamma(\nu)}f(s)\right]_{s=t}^{s=b} - \frac{1}{\Gamma(\nu)}\sum_{s=t}^{b-1}(\nu-1)(s+\nu-\sigma(t))^{(\nu-2)}f^\sigma(s)
$$
$$
= \frac{(b+\nu-\sigma(t))^{(\nu-1)}}{\Gamma(\nu)}f(b) - \frac{(\nu-1)^{(\nu-1)}}{\Gamma(\nu)}f(t)
$$
$$
\quad - \frac{1}{\Gamma(\nu)}\sum_{s=t}^{b-1}(\nu-1)(s+\nu-\sigma(t))^{(\nu-2)}f^\sigma(s).
$$

We now compute $\Delta({}_t\Delta_b^{-\nu}f(t))$:

$$
\Delta({}_t\Delta_b^{-\nu}f(t)) = \frac{1}{\Gamma(\nu)}\left[\sum_{s=\sigma(t)}^{b}(s+\nu-\sigma(t+1))^{(\nu-1)}f(s)\right.
$$
$$
\left. - \sum_{s=t}^{b}(s+\nu-\sigma(t))^{(\nu-1)}f(s)\right]
$$

$$= \frac{1}{\Gamma(\nu)} \left[\sum_{s=\sigma(t)}^{b} (s+\nu - \sigma(t+1))^{(\nu-1)} f(s) \right.$$

$$\left. - \sum_{s=\sigma(t)}^{b} (s+\nu - \sigma(t))^{(\nu-1)} f(s) \right] - \frac{(\nu-1)^{(\nu-1)}}{\Gamma(\nu)} f(t)$$

$$= \frac{1}{\Gamma(\nu)} \sum_{s=\sigma(t)}^{b} \Delta_t (s+\nu - \sigma(t))^{(\nu-1)} f(s) - \frac{(\nu-1)^{(\nu-1)}}{\Gamma(\nu)} f(t)$$

$$= -\frac{1}{\Gamma(\nu)} \sum_{s=t}^{b-1} (\nu-1)(s+\nu-\sigma(t))^{(\nu-2)} f^\sigma(s) - \frac{(\nu-1)^{(\nu-1)}}{\Gamma(\nu)} f(t).$$

Since t is arbitrary, the theorem is proved. $\square$

Definition 12.15. *Let $0 < \alpha \leq 1$ and set $\mu = 1 - \alpha$. Then, the left fractional difference and the right fractional difference of order α of a function $f \in \mathcal{F}$ are defined, respectively, by*

$$_a\Delta_t^\alpha f(t) = \Delta(_a\Delta_t^{-\mu} f(t)), \quad t \in \mathbb{T}^\kappa,$$

and

$$_t\Delta_b^\alpha f(t) = -\Delta(_t\Delta_b^{-\mu} f(t))), \quad t \in \mathbb{T}^\kappa.$$

12.5.3   *Fractional summation by parts*

Our aim is to introduce the discrete-time (in time scale $\mathbb{T} = \{a, a+1, \ldots, b\}$) fractional problem of the calculus of variations and to prove corresponding necessary optimality conditions. In order to obtain an analogue of the Euler–Lagrange equation (cf. Theorem 12.9) we first prove a fractional formula of summation by parts. The results give discrete analogues to the fractional Riemann–Liouville results available in the literature: Theorem 12.8 is the discrete analogue of fractional integration by parts (Riewe, 1996; Samko, Kilbas and Marichev, 1993); Theorem 12.9 is the discrete analogue of the fractional Euler–Lagrange equation of Agrawal (Agrawal, 2002, Theorem 1); the natural boundary conditions (12.46) and (12.47) are the discrete fractional analogues of the transversality conditions in (Agrawal, 2006; Almeida, Malinowska and Torres, 2010).

The next lemma is used in the proof of Theorem 12.8.

Lemma 12.5. *Let f and h be two functions defined on $\mathbb{T}^\kappa$ and g a function defined on $\mathbb{T}^\kappa \times \mathbb{T}^\kappa$. Then, the equality*

$$\sum_{\tau=a}^{b-1} f(\tau) \sum_{s=a}^{\tau-1} g(\tau,s)h(s) = \sum_{\tau=a}^{b-2} h(\tau) \sum_{s=\sigma(\tau)}^{b-1} g(s,\tau)f(s)$$

holds.

Proof. Choose $\mathbb{T} = \mathbb{Z}$ and $F(\tau,s) = f(\tau)g(\tau,s)h(s)$ in Theorem 10 of (Akin, 2002). □

The next result gives a fractional summation by parts formula. It was first proved in (Bastos, Ferreira and Torres, 2011a, Theorem 3.2).

Theorem 12.8 (Fractional summation by parts). *Let f and g be real valued functions defined on $\mathbb{T}^k$ and $\mathbb{T}$, respectively. Fix $0 < \alpha \le 1$ and put $\mu = 1 - \alpha$. Then,*

$$\sum_{t=a}^{b-1} f(t)\,_a\Delta_t^\alpha g(t) = f(b-1)g(b) - f(a)g(a) + \sum_{t=a}^{b-2} {}_t\Delta_{\rho(b)}^\alpha f(t)g^\sigma(t)$$

$$+ \frac{\mu\, g(a)}{\Gamma(\mu+1)} \left(\sum_{t=a}^{b-1} (t+\mu-a)^{(\mu-1)} f(t) - \sum_{t=\sigma(a)}^{b-1} (t+\mu-\sigma(a))^{(\mu-1)} f(t) \right).$$

Proof. From (12.29) we can write

$$\sum_{t=a}^{b-1} f(t)\,_a\Delta_t^\alpha g(t) = \sum_{t=a}^{b-1} f(t)\Delta({}_a\Delta_t^{-\mu} g(t))$$

$$= \sum_{t=a}^{b-1} f(t) \left[{}_a\Delta_t^{-\mu}\Delta g(t) + \frac{\mu}{\Gamma(\mu+1)}(t+\mu-a)^{(\mu-1)} g(a) \right]$$

$$= \sum_{t=a}^{b-1} f(t)\,_a\Delta_t^{-\mu}\Delta g(t) + \sum_{t=a}^{b-1} \frac{\mu}{\Gamma(\mu+1)}(t+\mu-a)^{(\mu-1)} f(t)g(a).$$

$$(12.33)$$

Using (12.28) we get

$$\sum_{t=a}^{b-1} f(t)\,_a\Delta_t^{-\mu}\Delta g(t)$$

$$= \sum_{t=a}^{b-1} f(t)\Delta g(t) + \frac{\mu}{\Gamma(\mu+1)} \sum_{t=a}^{b-1} f(t) \sum_{s=a}^{t-1} (t+\mu-\sigma(s))^{(\mu-1)}\Delta g(s)$$

$$= \sum_{t=a}^{b-1} f(t)\Delta g(t) + \frac{\mu}{\Gamma(\mu+1)} \sum_{t=a}^{b-2} \Delta g(t) \sum_{s=\sigma(t)}^{b-1} (s+\mu-\sigma(t))^{(\mu-1)} f(s)$$

$$= f(b-1)[g(b)-g(b-1)] + \sum_{t=a}^{b-2} \Delta g(t)\,_t\Delta_{\rho(b)}^{-\mu} f(t),$$

where the third equality follows by Lemma 12.5. We proceed to develop the right-hand side of the last equality as follows:

$$f(b-1)[g(b)-g(b-1)] + \sum_{t=a}^{b-2} \Delta g(t)\,_t\Delta_{\rho(b)}^{-\mu} f(t)$$

$$= f(b-1)[g(b)-g(b-1)] + \left[g(t)\,_t\Delta_{\rho(b)}^{-\mu} f(t)\right]_{t=a}^{t=b-1}$$

$$- \sum_{t=a}^{b-2} g^{\sigma}(t)\Delta(\,_t\Delta_{\rho(b)}^{-\mu} f(t))$$

$$= f(b-1)g(b) - f(a)g(a) - \frac{\mu}{\Gamma(\mu+1)}g(a) \sum_{s=\sigma(a)}^{b-1} (s+\mu-\sigma(a))^{(\mu-1)} f(s)$$

$$+ \sum_{t=a}^{b-2} \left(\,_t\Delta_{\rho(b)}^{\alpha} f(t)\right) g^{\sigma}(t),$$

where the first equality follows from the usual summation by parts formula. Putting this into (12.33), we get:

$$\sum_{t=a}^{b-1} f(t)\,_a\Delta_t^{\alpha} g(t) = f(b-1)g(b) - f(a)g(a) + \sum_{t=a}^{b-2} \left(\,_t\Delta_{\rho(b)}^{\alpha} f(t)\right) g^{\sigma}(t)$$

$$+ \frac{g(a)\mu}{\Gamma(\mu+1)} \sum_{t=a}^{b-1} \frac{(t+\mu-a)^{(\mu-1)}}{\Gamma(\mu)} f(t) - \frac{g(a)\mu}{\Gamma(\mu+1)} \sum_{s=\sigma(a)}^{b-1} (s+\mu-\sigma(a))^{(\mu-1)} f(s).$$

The theorem is proved. □

12.5.4 *Necessary optimality conditions*

We begin to fix two arbitrary real numbers α and β such that $\alpha, \beta \in (0, 1]$. Further, we put $\mu = 1 - \alpha$ and $\nu = 1 - \beta$.

Let a function $L(t, u, v, w) : \mathbb{T}^\kappa \times \mathbb{R} \times \mathbb{R} \times \mathbb{R} \to \mathbb{R}$ be given. We assume that the second-order partial derivatives L_{uu}, L_{uv}, L_{uw}, L_{vw}, L_{vv}, and L_{ww} exist and are continuous.

Consider the functional $\mathcal{L} : \mathcal{F} \to \mathbb{R}$ defined by

$$\mathcal{L}[y] = \sum_{t=a}^{b-1} L(t, y^\sigma(t), {}_a\Delta_t^\alpha y(t), {}_t\Delta_b^\beta y(t)) \tag{12.34}$$

and the problem, that we denote by (P), of minimizing (12.34) subject to the boundary conditions $y(a) = A$ and $y(b) = B$ ($A, B \in \mathbb{R}$). Our aim is to derive necessary conditions of first and second-order for problem (P).

Definition 12.16. *For $f \in \mathcal{F}$ we define the norm*

$$\|f\| = \max_{t\in\mathbb{T}^\kappa} |f^\sigma(t)| + \max_{t\in\mathbb{T}^\kappa} |{}_a\Delta_t^\alpha f(t)| + \max_{t\in\mathbb{T}^\kappa} |{}_t\Delta_b^\beta f(t)|.$$

A function $\tilde{y} \in \mathcal{F}$ with $\tilde{y}(a) = A$ and $\tilde{y}(b) = B$ is called a local minimizer for problem (P) provided there exists $\delta > 0$ such that $\mathcal{L}[\tilde{y}] \leq \mathcal{L}[y]$ for all $y \in \mathcal{F}$ with $y(a) = A$ and $y(b) = B$ and $\|y - \tilde{y}\| < \delta$.

Remark 12.14. It is easy to see that Definition 12.16 gives a norm in $\mathcal{F}$. Indeed, it is clear that $\|f\|$ is non-negative, and for an arbitrary $f \in \mathcal{F}$ and $k \in \mathbb{R}$ we have $\|kf\| = |k|\|f\|$. The triangle inequality is also easy to prove:

$$\|f + g\| = \max_{t\in\mathbb{T}^\kappa} |f(t) + g(t)| + \max_{t\in\mathbb{T}^\kappa} |{}_a\Delta_t^\alpha(f + g)(t)| + \max_{t\in\mathbb{T}^\kappa} |{}_t\Delta_b^\alpha(f + g)(t)|$$

$$\leq \max_{t\in\mathbb{T}^\kappa} [|f(t)| + |g(t)|] + \max_{t\in\mathbb{T}^\kappa} [|{}_a\Delta_t^\alpha f(t)| + |{}_a\Delta_t^\alpha g(t)|]$$

$$+ \max_{t\in\mathbb{T}^\kappa} [|{}_t\Delta_b^\alpha f(t)| + |{}_t\Delta_b^\alpha g(t)|]$$

$$\leq \|f\| + \|g\|.$$

The only possible doubt is to prove that $\|f\| = 0$ implies that $f(t) = 0$ for any $t \in \mathbb{T} = \{a, a+1, \ldots, b\}$. Suppose $\|f\| = 0$. It follows that

$$\max_{t\in\mathbb{T}^\kappa} |f^\sigma(t)| = 0, \tag{12.35}$$

$$\max_{t\in\mathbb{T}^\kappa} |{}_a\Delta_t^\alpha f(t)| = 0, \tag{12.36}$$

$$\max_{t\in\mathbb{T}^\kappa} |{}_t\Delta_b^\beta f(t)| = 0. \tag{12.37}$$

From (12.35) we conclude that $f(t) = 0$ for all $t \in \{a+1, \ldots, b\}$. It remains to prove that $f(a) = 0$. To prove this we use (12.36) (or (12.37)). Indeed, from (12.35) we can write

$$
_a\Delta_t^\alpha f(t) = \Delta \left(\frac{1}{\Gamma(1-\alpha)} \sum_{s=a}^{t} (t+1-\alpha-\sigma(s))^{(-\alpha)} f(s) \right)
$$

$$
= \frac{\displaystyle\sum_{s=a}^{t+1} (t+2-\alpha-\sigma(s))^{(-\alpha)} f(s) - \sum_{s=a}^{t} (t+1-\alpha-\sigma(s))^{(-\alpha)} f(s)}{\Gamma(1-\alpha)}
$$

$$
= \frac{(t+2-\alpha-\sigma(a))^{(-\alpha)} f(a) - (t+1-\alpha-\sigma(a))^{(-\alpha)} f(a)}{\Gamma(1-\alpha)}
$$

$$
= \frac{f(a)}{\Gamma(1-\alpha)} \Delta(t+1-\alpha-\sigma(a))^{(-\alpha)}
$$

and since by (12.36) $_a\Delta_t^\alpha f(t) = 0$, one concludes that $f(a) = 0$ (because $(t+1-\alpha-\sigma(a))^{(-\alpha)}$ is not a constant).

Definition 12.17. *A function $\eta \in \mathcal{F}$ is called an admissible variation for problem (P) provided $\eta \neq 0$ and $\eta(a) = \eta(b) = 0$.*

The next theorem presents a first-order necessary optimality condition for problem (P). It was first proved in (Bastos, Ferreira and Torres, 2011a, Theorem 3.5).

Theorem 12.9 (Fractional discrete-time Euler–Lagrange equation).
If $\tilde{y} \in \mathcal{F}$ is a local minimizer for problem (P), then

$$
L_u[\tilde{y}](t) + {}_t\Delta_{\rho(b)}^\alpha L_v[\tilde{y}](t) + {}_a\Delta_t^\beta L_w[\tilde{y}](t) = 0 \tag{12.38}
$$

holds for all $t \in \mathbb{T}^{\kappa^2}$, where the operator $[\cdot]$ is defined by

$$
[y](s) = (s, y^\sigma(s), {}_a\Delta_s^\alpha y(s), {}_s\Delta_b^\beta y(s)).
$$

Proof. Suppose that $\tilde{y}$ is a local minimizer of $\mathcal{L}$. Let η be an arbitrary fixed admissible variation and define the function $\Phi : \left(-\frac{\delta}{\|\eta\|}, \frac{\delta}{\|\eta\|} \right) \to \mathbb{R}$ by

$$
\Phi(\varepsilon) = \mathcal{L}[\tilde{y} + \varepsilon\eta]. \tag{12.39}
$$

This function has a minimum at $\varepsilon = 0$, so we must have $\Phi'(0) = 0$, i.e.,

$$
\sum_{t=a}^{b-1} \left[L_u[\tilde{y}](t)\eta^\sigma(t) + L_v[\tilde{y}](t){}_a\Delta_t^\alpha\eta(t) + L_w[\tilde{y}](t){}_t\Delta_b^\beta\eta(t) \right] = 0,
$$

which we may write, equivalently, as

$$L_u[\tilde{y}](t)\eta^\sigma(t)|_{t=\rho(b)} + \sum_{t=a}^{b-2} L_u[\tilde{y}](t)\eta^\sigma(t)$$

$$+ \sum_{t=a}^{b-1} L_v[\tilde{y}](t){}_a\Delta_t^\alpha \eta(t) + \sum_{t=a}^{b-1} L_w[\tilde{y}](t){}_t\Delta_b^\beta \eta(t) = 0. \quad (12.40)$$

Using Theorem 12.8, and the fact that $\eta(a) = \eta(b) = 0$, we get for the third term in (12.40) that

$$\sum_{t=a}^{b-1} L_v[\tilde{y}](t){}_a\Delta_t^\alpha \eta(t) = \sum_{t=a}^{b-2} \left({}_t\Delta_{\rho(b)}^\alpha L_v[\tilde{y}](t)\right)\eta^\sigma(t). \quad (12.41)$$

Using (12.30) it follows that

$$\sum_{t=a}^{b-1} L_w[\tilde{y}](t){}_t\Delta_b^\beta \eta(t)$$

$$= -\sum_{t=a}^{b-1} L_w[\tilde{y}](t)\Delta({}_t\Delta_b^{-\nu}\eta(t))$$

$$= -\sum_{t=a}^{b-1} L_w[\tilde{y}](t)\left[{}_t\Delta_{\rho(b)}^{-\nu}\Delta\eta(t) - \frac{\nu}{\Gamma(\nu+1)}(b+\nu-\sigma(t))^{(\nu-1)}\eta(b)\right]$$

$$= -\left(\sum_{t=a}^{b-1} L_w[\tilde{y}](t){}_t\Delta_{\rho(b)}^{-\nu}\Delta\eta(t) - \frac{\nu\eta(b)}{\Gamma(\nu+1)}\sum_{t=a}^{b-1}(b+\nu-\sigma(t))^{(\nu-1)}L_w[\tilde{y}](t)\right).$$

$$(12.42)$$

We now use Lemma 12.5 to get

$$\sum_{t=a}^{b-1} L_w[\tilde{y}](t){}_t\Delta_{\rho(b)}^{-\nu}\Delta\eta(t)$$

$$= \sum_{t=a}^{b-1} L_w[\tilde{y}](t)\Delta\eta(t) + \frac{\nu}{\Gamma(\nu+1)}\sum_{t=a}^{b-2} L_w[\tilde{y}](t)\sum_{s=\sigma(t)}^{b-1}(s+\nu-\sigma(t))^{(\nu-1)}\Delta\eta(s)$$

$$= \sum_{t=a}^{b-1} L_w[\tilde{y}](t)\Delta\eta(t) + \frac{\nu}{\Gamma(\nu+1)}\sum_{t=a}^{b-1}\Delta\eta(t)\sum_{s=a}^{t-1}(t+\nu-\sigma(s))^{(\nu-1)}L_w[\tilde{y}](s)$$

$$= \sum_{t=a}^{b-1}\Delta\eta(t){}_a\Delta_t^{-\nu}L_w[\tilde{y}](t).$$

$$(12.43)$$

We apply again the usual summation by parts formula, this time to (12.43), to obtain:

$$\sum_{t=a}^{b-1} \Delta\eta(t)_a\Delta_t^{-\nu}L_w[\tilde{y}](t)$$

$$= \sum_{t=a}^{b-2} \Delta\eta(t)_a\Delta_t^{-\nu}L_w[\tilde{y}](t) + (\eta(b) - \eta(\rho(b)))_a\Delta_t^{-\nu}L_w[\tilde{y}](t)|_{t=\rho(b)}$$

$$= \left[\eta(t)_a\Delta_t^{-\nu}L_w[\tilde{y}](t)\right]_{t=a}^{t=b-1} - \sum_{t=a}^{b-2} \eta^\sigma(t)\Delta(_a\Delta_t^{-\nu}L_w[\tilde{y}](t))$$

$$+ \eta(b)_a\Delta_t^{-\nu}L_w[\tilde{y}](t)|_{t=\rho(b)} - \eta(b-1)_a\Delta_t^{-\nu}L_w[\tilde{y}](t)|_{t=\rho(b)}$$

$$= \eta(b)_a\Delta_t^{-\nu}L_w[\tilde{y}](t)|_{t=\rho(b)} - \eta(a)_a\Delta_t^{-\nu}L_w[\tilde{y}](t)|_{t=a}$$

$$- \sum_{t=a}^{b-2} \eta^\sigma(t)_a\Delta_t^\beta L_w[\tilde{y}](t).$$

$$(12.44)$$

Since $\eta(a) = \eta(b) = 0$ it follows, from (12.43) and (12.44), that

$$\sum_{t=a}^{b-1} L_w[\tilde{y}](t)_t\Delta_{\rho(b)}^{-\nu}\Delta\eta(t) = -\sum_{t=a}^{b-2} \eta^\sigma(t)_a\Delta_t^\beta L_w[\tilde{y}](t)$$

and, after inserting in (12.42), that

$$\sum_{t=a}^{b-1} L_w[\tilde{y}](t)_t\Delta_b^\beta\eta(t) = \sum_{t=a}^{b-2} \eta^\sigma(t)_a\Delta_t^\beta L_w[\tilde{y}](t). \qquad (12.45)$$

By (12.41) and (12.45) we may write (12.40) as

$$\sum_{t=a}^{b-2} \left[L_u[\tilde{y}](t) + {}_t\Delta_{\rho(b)}^\alpha L_v[\tilde{y}](t) + {}_a\Delta_t^\beta L_w[\tilde{y}](t)\right]\eta^\sigma(t) = 0.$$

Since the values of $\eta^\sigma(t)$ are arbitrary for $t \in \mathbb{T}^{\kappa^2}$, the Euler–Lagrange equation (12.38) holds along $\tilde{y}$. $\square$

Remark 12.15. If the initial condition $y(a) = A$ is not present (i.e., $y(a)$ is free), we can use standard techniques to show that the following supplementary condition must be fulfilled:

$$- L_v(a) + \frac{\mu}{\Gamma(\mu+1)}\left(\sum_{t=a}^{b-1}(t+\mu-a)^{(\mu-1)}L_v[\tilde{y}](t)\right.$$

$$\left. - \sum_{t=\sigma(a)}^{b-1}(t+\mu-\sigma(a))^{(\mu-1)}L_v[\tilde{y}](t)\right) + L_w(a) = 0. \quad (12.46)$$

Similarly, if $y(b) = B$ is not present (i.e., $y(b)$ is free), the equality

$$L_u(\rho(b)) + L_v(\rho(b)) - L_w(\rho(b))$$

$$+ \frac{\nu}{\Gamma(\nu+1)} \left(\sum_{t=a}^{b-1} (b + \nu - \sigma(t))^{(\nu-1)} L_w[\tilde{y}](t) \right.$$

$$\left. - \sum_{t=a}^{b-2} (\rho(b) + \nu - \sigma(t))^{(\nu-1)} L_w[\tilde{y}](t) \right) = 0 \quad (12.47)$$

holds. We just note that the first term in (12.47) arises from the first term on the left-hand side of (12.40). Equalities (12.46) and (12.47) are the fractional discrete-time natural boundary conditions.

The next result is a particular case of Theorem 12.9 and can be found, e.g., in (Bohner, 2004; Ferreira and Torres, 2008).

Corollary 12.1 (The discrete-time Euler–Lagrange equation).
If $\tilde{y}$ is a solution to the problem

$$\mathcal{L}[y] = \sum_{t=a}^{b-1} L(t, y(t+1), \Delta y(t)) \longrightarrow \min \tag{12.48}$$

$$y(a) = A, \quad y(b) = B,$$

then

$$L_u(t, \tilde{y}(t+1), \Delta \tilde{y}(t)) - \Delta L_v(t, \tilde{y}(t+1), \Delta \tilde{y}(t)) = 0$$

for all $t \in \{a, \ldots, b-2\}$.

Proof. Follows from Theorem 12.9 with $\alpha = 1$ and a L not depending on w. $\qquad\square$

We derive now the second-order necessary condition for problem (P), that is, we obtain Legendre's necessary condition for the fractional difference setting. This result was first proved in (Bastos, Ferreira and Torres, 2011a, Theorem 3.6).

Theorem 12.10 (Fractional discrete-time Legendre condition).
If $\tilde{y} \in \mathcal{F}$ is a local minimizer for problem (P), then the inequality

$$L_{uu}[\tilde{y}](t) + 2L_{uv}[\tilde{y}](t) + L_{vv}[\tilde{y}](t) + L_{vv}[\tilde{y}](\sigma(t))(\mu - 1)^2$$

$$+ \sum_{s=\sigma(\sigma(t))}^{b-1} L_{vv}[\tilde{y}](s) \left(\frac{\mu(\mu-1)\prod_{i=0}^{s-t-3}(\mu+i+1)}{(s-t)\Gamma(s-t)} \right)^2 + 2L_{uw}[\tilde{y}](t)(\nu - 1)$$

$$+ 2(\nu - 1)L_{vw}[\tilde{y}](t) + 2(\mu - 1)L_{vw}[\tilde{y}](\sigma(t)) + L_{ww}[\tilde{y}](t)(1 - \nu)^2$$

$$+ L_{ww}[\tilde{y}](\sigma(t)) + \sum_{s=a}^{t-1} L_{ww}[\tilde{y}](s) \left(\frac{\nu(1-\nu)\prod_{i=0}^{t-s-2}(\nu+i)}{(\sigma(t)-s)\Gamma(\sigma(t)-s)} \right)^2 \geq 0$$

holds for all $t \in \mathbb{T}^{\kappa^2}$, where $[\tilde{y}](t) = (t, \tilde{y}^\sigma(t), {}_a\Delta_t^\alpha \tilde{y}(t), {}_t\Delta_b^\beta \tilde{y}(t))$.

Proof.   By the hypothesis of the theorem, and letting Φ be as in (12.39), we get

$$\Phi''(0) \geq 0 \tag{12.49}$$

for an arbitrary admissible variation η. Inequality (12.49) is equivalent to

$$\sum_{t=a}^{b-1} \Big[L_{uu}[\tilde{y}](t)(\eta^\sigma(t))^2 + 2L_{uv}[\tilde{y}](t)\eta^\sigma(t){}_a\Delta_t^\alpha \eta(t) + L_{vv}[\tilde{y}](t)({}_a\Delta_t^\alpha \eta(t))^2$$

$$+ 2L_{uw}[\tilde{y}](t)\eta^\sigma(t){}_t\Delta_b^\beta \eta(t) + 2L_{vw}[\tilde{y}](t){}_a\Delta_t^\alpha \eta(t){}_t\Delta_b^\beta \eta(t)$$

$$+ L_{ww}[\tilde{y}](t)({}_t\Delta_b^\beta \eta(t))^2 \Big] \geq 0.$$

Let $\tau \in \mathbb{T}^{\kappa^2}$ be arbitrary and define $\eta : \mathbb{T} \to \mathbb{R}$ by

$$\eta(t) = \begin{cases} 1 \text{ if } t = \sigma(\tau); \\ 0 \text{ otherwise.} \end{cases}$$

It follows that $\eta(a) = \eta(b) = 0$, that is, η is an admissible variation

Using (12.29) (note that $\eta(a) = 0$), we get

$$\sum_{t=a}^{b-1} \left[L_{uu}[\tilde{y}](t)(\eta^\sigma(t))^2 + 2L_{uv}[\tilde{y}](t)\eta^\sigma(t) \,_a\Delta_t^\alpha \eta(t) + L_{vv}[\tilde{y}](t)(\,_a\Delta_t^\alpha \eta(t))^2 \right]$$

$$= \sum_{t=a}^{b-1} \left\{ L_{uu}[\tilde{y}](t)(\eta^\sigma(t))^2 \right.$$

$$+ 2L_{uv}[\tilde{y}](t)\eta^\sigma(t) \left[\Delta\eta(t) + \frac{\mu}{\Gamma(\mu+1)} \sum_{s=a}^{t-1} (t + \mu - \sigma(s))^{(\mu-1)} \Delta\eta(s) \right]$$

$$\left. + L_{vv}[\tilde{y}](t) \left(\Delta\eta(t) + \frac{\mu}{\Gamma(\mu+1)} \sum_{s=a}^{t-1} (t + \mu - \sigma(s))^{(\mu-1)} \Delta\eta(s) \right)^2 \right\}$$

$$= L_{uu}[\tilde{y}](\tau) + 2L_{uv}[\tilde{y}](\tau) + L_{vv}[\tilde{y}](\tau)$$

$$+ \sum_{t=\sigma(\tau)}^{b-1} L_{vv}[\tilde{y}](t) \left(\Delta\eta(t) + \frac{\mu}{\Gamma(\mu+1)} \sum_{s=a}^{t-1} (t + \mu - \sigma(s))^{(\mu-1)} \Delta\eta(s) \right)^2.$$

Observe that

$$L_{vv}(\sigma(\tau))(-1 + \mu)^2$$

$$+ \sum_{t=\sigma(\sigma(\tau))}^{b-1} L_{vv}[\tilde{y}](t) \left(\frac{\mu}{\Gamma(\mu+1)} \sum_{s=a}^{t-1} (t + \mu - \sigma(s))^{(\mu-1)} \Delta\eta(s) \right)^2$$

$$= \sum_{t=\sigma(\tau)}^{b-1} L_{vv}[\tilde{y}](t) \left(\Delta\eta(t) + \frac{\mu}{\Gamma(\mu+1)} \sum_{s=a}^{t-1} (t + \mu - \sigma(s))^{(\mu-1)} \Delta\eta(s) \right)^2.$$

We show next that

$$\sum_{t=\sigma(\sigma(\tau))}^{b-1} L_{vv}[\tilde{y}](t) \left(\frac{\mu}{\Gamma(\mu+1)} \sum_{s=a}^{t-1} (t + \mu - \sigma(s))^{(\mu-1)} \Delta\eta(s) \right)^2$$

$$= \sum_{t=\sigma(\sigma(\tau))}^{b-1} L_{vv}[\tilde{y}](t) \left(\frac{\mu(\mu-1) \prod_{i=0}^{t-\tau-3}(\mu + i + 1)}{(t-\tau)\Gamma(t-\tau)} \right)^2.$$

Let $t \in [\sigma(\sigma(\tau)), b - 1] \cap \mathbb{Z}$. Then,

$$\frac{\mu}{\Gamma(\mu + 1)} \sum_{s=a}^{t-1} (t + \mu - \sigma(s))^{(\mu-1)} \Delta \eta(s)$$

$$= \frac{\mu}{\Gamma(\mu + 1)} \left[\sum_{s=a}^{\tau} (t + \mu - \sigma(s))^{(\mu-1)} \Delta \eta(s) \right.$$

$$\left. + \sum_{s=\sigma(\tau)}^{t-1} (t + \mu - \sigma(s))^{(\mu-1)} \Delta \eta(s) \right]$$

$$= \frac{\mu}{\Gamma(\mu + 1)} \left[(t + \mu - \sigma(\tau))^{(\mu-1)} - (t + \mu - \sigma(\sigma(\tau)))^{(\mu-1)} \right]$$

$$= \frac{\mu}{\Gamma(\mu + 1)} \left[\frac{\Gamma(t + \mu - \sigma(\tau) + 1)}{\Gamma(t + \mu - \sigma(\tau) + 1 - (\mu - 1))} \right.$$

$$\left. = -\frac{\Gamma(t + \mu - \sigma(\sigma(\tau)) + 1)}{\Gamma(t + \mu - \sigma(\sigma(\tau)) + 1 - (\mu - 1))} \right]$$

$$= \frac{\mu}{\Gamma(\mu + 1)} \left[\frac{\Gamma(t + \mu - \tau)}{\Gamma(t - \tau + 1)} - \frac{\Gamma(t - \tau + \mu - 1)}{\Gamma(t - \tau)} \right]$$

$$= \frac{\mu}{\Gamma(\mu + 1)} \left[\frac{(t + \mu - \tau - 1)\Gamma(t + \mu - \tau - 1)}{(t - \tau)\Gamma(t - \tau)} - \frac{(t - \tau)\Gamma(t - \tau + \mu - 1)}{(t - \tau)\Gamma(t - \tau)} \right]$$

$$= \frac{\mu}{\Gamma(\mu + 1)} \frac{(\mu - 1)\Gamma(t - \tau + \mu - 1)}{(t - \tau)\Gamma(t - \tau)}$$

$$= \frac{\mu(\mu - 1) \prod_{i=0}^{t-\tau-3}(\mu + i + 1)}{(t - \tau)\Gamma(t - \tau)},$$

$$(12.50)$$

which proves our claim. Observe that we can write

$$_t\Delta_b^{\beta} \eta(t) = -_t\Delta_{\rho(b)}^{-\nu} \Delta \eta(t)$$

since $\eta(b) = 0$. It is not difficult to see that the following equality holds:

$$\sum_{t=a}^{b-1} 2L_{uw}[\tilde{y}](t)\eta^{\sigma}(t)_t\Delta_b^{\beta} \eta(t)$$

$$= -\sum_{t=a}^{b-1} 2L_{uw}[\tilde{y}](t)\eta^{\sigma}(t)_t\Delta_{\rho(b)}^{-\nu} \Delta \eta(t) = 2L_{uw}[\tilde{y}](\tau)(\nu - 1).$$

Moreover,

$$\sum_{t=a}^{b-1} 2L_{vw}[\tilde{y}](t)\,_a\Delta_t^\alpha \eta(t)\,_t\Delta_b^\beta \eta(t)$$

$$= -2\sum_{t=a}^{b-1} L_{vw}[\tilde{y}](t)\left\{\left(\Delta\eta(t) + \frac{\mu}{\Gamma(\mu+1)}\cdot\sum_{s=a}^{t-1}(t+\mu-\sigma(s))^{(\mu-1)}\Delta\eta(s)\right)\right.$$

$$\left.\cdot\left[\Delta\eta(t) + \frac{\nu}{\Gamma(\nu+1)}\sum_{s=\sigma(t)}^{b-1}(s+\nu-\sigma(t))^{(\nu-1)}\Delta\eta(s)\right]\right\}$$

$$= 2(\nu-1)L_{vw}[\tilde{y}](\tau) + 2(\mu-1)L_{vw}[\tilde{y}](\sigma(\tau)).$$

Finally, we have that

$$\sum_{t=a}^{b-1} L_{ww}[\tilde{y}](t)(\,_t\Delta_b^\beta \eta(t))^2$$

$$= \sum_{t=a}^{\sigma(\tau)} L_{ww}[\tilde{y}](t)\left[\Delta\eta(t) + \frac{\nu}{\Gamma(\nu+1)}\sum_{s=\sigma(t)}^{b-1}(s+\nu-\sigma(t))^{(\nu-1)}\Delta\eta(s)\right]^2$$

$$= \sum_{t=a}^{\tau-1} L_{ww}[\tilde{y}](t)\left[\frac{\nu}{\Gamma(\nu+1)}\sum_{s=\sigma(t)}^{b-1}(s+\nu-\sigma(t))^{(\nu-1)}\Delta\eta(s)\right]^2$$

$$+ L_{ww}[\tilde{y}](\tau)(1-\nu)^2 + L_{ww}[\tilde{y}](\sigma(\tau))$$

$$= \sum_{t=a}^{\tau-1} L_{ww}[\tilde{y}](t)\left[\frac{\nu}{\Gamma(\nu+1)}\left\{(\tau+\nu-\sigma(t))^{(\nu-1)} - (\sigma(\tau)+\nu-\sigma(t))^{(\nu-1)}\right\}\right]^2$$

$$+ L_{ww}[\tilde{y}](\tau)(1-\nu)^2 + L_{ww}[\tilde{y}](\sigma(\tau)).$$

Similarly as we have done in (12.50), we obtain that

$$\frac{\nu}{\Gamma(\nu+1)}\left[(\tau+\nu-\sigma(t))^{(\nu-1)} - (\sigma(\tau)+\nu-\sigma(t))^{(\nu-1)}\right]$$

$$= \frac{\nu(1-\nu)\prod_{i=0}^{\tau-t-2}(\nu+i)}{(\sigma(\tau)-t)\Gamma(\sigma(\tau)-t)}.$$

We are done with the proof. □

A trivial corollary gives the discrete-time version of Legendre's necessary condition (cf., e.g., (Bohner, 2004; Hilscher and Zeidan, 2003)).

Corollary 12.2 (The discrete-time Legendre condition). *If $\tilde{y}$ is a solution to the problem* (12.48), *then*

$$L_{uu}[\tilde{y}](t) + 2L_{uv}[\tilde{y}](t) + L_{vv}[\tilde{y}](t) + L_{vv}[\tilde{y}](\sigma(t)) \geq 0$$

holds for all $t \in \mathbb{T}^{\kappa^2}$, where $[\tilde{y}](t) = (t, \tilde{y}^{\sigma}(t), \Delta\tilde{y}(t))$.

Proof. We consider problem (P) with $\alpha = 1$ and L not depending on w. The choice $\alpha = 1$ implies $\mu = 0$, and the result follows immediately from Theorem 12.10. $\qquad\square$

12.5.5 *Examples*

In this section we present three illustrative examples. The results were obtained using the open source Computer Algebra System Maxima.[1] All computations were done running Maxima on an Intel® Core^{TM}2 Duo, CPU of 2.27GHz with 3Gb of RAM.

Example 12.9. Let us consider the following problem:

$$J_\alpha[y] = \sum_{t=0}^{b-1} \left({}_0\Delta_t^\alpha y(t)\right)^2 \longrightarrow \min, \quad y(0) = A, \quad y(b) = B. \qquad (12.51)$$

In this case Theorem 12.10 is trivially satisfied. We obtain the solution $\tilde{y}$ to the Euler–Lagrange equation (12.38) for the case $b = 2$ using the computer algebra system Maxima. Using the Maxima package (see the definition of the command `extremal` in (Bastos, 2012) or in (Bastos, Ferreira and Torres, 2011a, Appendix)) we do

```
L1:v^2$
extremal(L1,0,2,A,B,alpha,alpha);
```

to obtain (2 seconds)

$$\tilde{y}(1) = \frac{2\,\alpha\,B + \left(\alpha^3 - \alpha^2 + 2\,\alpha\right)A}{2\,\alpha^2 + 2}. \qquad (12.52)$$

For the particular case $\alpha = 1$ the equality (12.52) gives $\tilde{y}(1) = \frac{A+B}{2}$, which coincides with the solution to the (non-fractional) discrete problem

$$\sum_{t=0}^{1} (\Delta y(t))^2 = \sum_{t=0}^{1} (y(t+1) - y(t))^2 \longrightarrow \min, \quad y(0) = A, \quad y(2) = B.$$

---

[1]http://maxima.sourceforge.net

Similarly, we can obtain exact formulas of the extremal on bigger intervals (for bigger values of b). For example, the solution of problem (12.51) with $b = 3$ is (35 seconds)

$$\tilde{y}(1) = \frac{(6\alpha^2 + 6\alpha)\,B + (2\alpha^5 + 2\alpha^4 + 10\alpha^3 - 2\alpha^2 + 12\alpha)\,A}{3\alpha^4 + 6\alpha^3 + 15\alpha^2 + 12},$$

$$\tilde{y}(2) = \frac{(12\alpha^3 + 12\alpha^2 + 24\alpha)\,B + (\alpha^6 + \alpha^5 + 7\alpha^4 - \alpha^3 + 4\alpha^2 + 12\alpha)\,A}{6\alpha^4 + 12\alpha^3 + 30\alpha^2 + 24};$$

and the solution of problem (12.51) with $b = 4$ is (72 seconds)

$$\tilde{y}(1) = \frac{3\alpha^7 + 15\alpha^6 + 57\alpha^5 + 69\alpha^4 + 156\alpha^3 - 12\alpha^2 + 144\alpha}{\xi}A$$
$$+ \frac{24\alpha^3 + 72\alpha^2 + 48\alpha}{\xi}B,$$

$$\tilde{y}(2) = \frac{\alpha^8 + 5\alpha^7 + 22\alpha^6 + 32\alpha^5 + 67\alpha^4 + 35\alpha^3 + 54\alpha^2 + 72\alpha}{\xi}A$$
$$+ \frac{24\alpha^4 + 72\alpha^3 + 120\alpha^2 + 72\alpha}{\xi}B,$$

$$\tilde{y}(3) = \frac{\alpha^9 + 6\alpha^8 + 30\alpha^7 + 60\alpha^6 + 117\alpha^5 + 150\alpha^4 - 4\alpha^3 + 216\alpha^2 + 288\alpha}{\zeta}A$$
$$+ \frac{72\alpha^5 + 288\alpha^4 + 792\alpha^3 + 576\alpha^2 + 864\alpha}{\zeta}B,$$

where

$$\xi = 4\alpha^6 + 24\alpha^5 + 88\alpha^4 + 120\alpha^3 + 196\alpha^2 + 144,$$
$$\zeta = 24\alpha^6 + 144\alpha^5 + 528\alpha^4 + 720\alpha^3 + 1176\alpha^2 + 864.$$

Consider now problem (12.51) with $b = 4$, $A = 0$, and $B = 1$. In Table 12.1 we show the extremal values $\tilde{y}(1)$, $\tilde{y}(2)$, $\tilde{y}(3)$, and corresponding $\tilde{J}_\alpha$, for some values of α. The numerical results show that the fractional extremal converges to the classical (integer-order) extremal when α tends to one. This is illustrated in Fig. 12.1. The numerical results from Table 12.1 and Fig. 12.2 show that for this problem the smallest value of $\tilde{J}_\alpha$, $\alpha \in]0, 1]$, occur for $\alpha = 1$ (i.e., the smallest value of $\tilde{J}_\alpha$ occurs for the classical non-fractional case).

Example 12.10. In this example we generalize problem (12.51) to

$$J_{\alpha,\beta} = \sum_{t=0}^{b-1} \gamma_1 \Big({}_0\Delta_t^\alpha y(t)\Big)^2 + \gamma_2 \Big({}_t\Delta_b^\beta y(t)\Big)^2 \longrightarrow \min, \tag{12.53}$$

$$y(0) = A, \quad y(b) = B.$$

As before, we solve the associated Euler–Lagrange equation (12.38) for the case $b = 2$ with the help of the Maxima package (35 seconds):

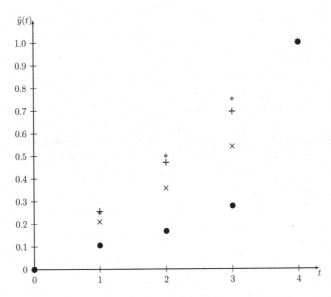

Fig. 12.1   Extremal $\tilde{y}(t)$ of Example 12.9 with $b = 4$, $A = 0$, $B = 1$, and different α's ($\bullet$: $\alpha = 0.25$; $\times$: $\alpha = 0.5$; $+$: $\alpha = 0.75$; $*$: $\alpha = 1$).

Table 12.1   The extremal values $\tilde{y}(1)$, $\tilde{y}(2)$ and $\tilde{y}(3)$ of problem (12.51) with $b = 4$, $A = 0$, and $B = 1$ for different α's.

α	$\tilde{y}(1)$	$\tilde{y}(2)$	$\tilde{y}(3)$	$\tilde{J}_\alpha$
0.25	0.10647146897355	0.16857982587479	0.2792657904952	0.90855653524095
0.50	0.20997375328084	0.35695538057743	0.54068241469816	0.67191601049869
0.75	0.25543605027861	0.4702345471038	0.69508876506414	0.4246209666969
1	0.25	0.5	0.75	0.25

```
L2:(gamma[1])*v^2+(gamma[2])*w^2$
extremal(L2,0,2,A,B,alpha,beta);
```

$$\tilde{y}(1) = \frac{\left(2\,\gamma_2\,\beta + \gamma_1\,\alpha^3 - \gamma_1\,\alpha^2 + 2\,\gamma_1\,\alpha\right)A}{2\,\gamma_2\,\beta^2 + 2\,\gamma_1\,\alpha^2 + 2\,\gamma_2 + 2\,\gamma_1}$$
$$+\frac{\left(\gamma_2\,\beta^3 - \gamma_2\,\beta^2 + 2\,\gamma_2\,\beta + 2\,\gamma_1\,\alpha\right)B}{2\,\gamma_2\,\beta^2 + 2\,\gamma_1\,\alpha^2 + 2\,\gamma_2 + 2\,\gamma_1}.$$

Consider now problem (12.53) with $\gamma_1 = \gamma_2 = 1$, $b = 2$, $A = 0$, $B = 1$, and $\beta = \alpha$. In Table 12.2 we show the values of $\tilde{y}(1)$ and $\tilde{J}_\alpha := J_{\alpha,\alpha}(\tilde{y}(1))$ for some values of α. We concluded, numerically, that the fractional extremal $\tilde{y}(1)$ tends to the classical (non-fractional) extremal when α tends to one.

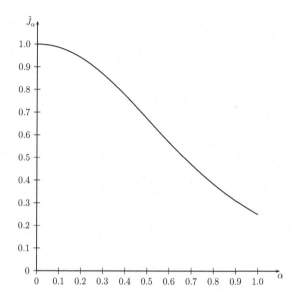

Fig. 12.2   Function $\tilde{J}_\alpha$ of Example 12.9 with $b = 4$, $A = 0$, and $B = 1$.

Differently from Example 12.9, the smallest value of $\tilde{J}_\alpha$, $\alpha \in]0, 1]$, does not occur here for $\alpha = 1$ (see Fig. 12.3). The smallest value of $\tilde{J}_\alpha$, $\alpha \in]0, 1]$, occurs for $\alpha = 0.61747447161482$.

Table 12.2   The extremal $\tilde{y}(1)$ of problem (12.53) for different values of α ($\gamma_1 = \gamma_2 = 1$, $b = 2$, $A = 0$, $B = 1$, and $\beta = \alpha$).

α	$\tilde{y}(1)$	$\tilde{J}_\alpha$
0.25	0.22426470588235	0.96441291360294
0.50	0.375	0.9140625
0.75	0.4575	0.91720703125
1	0.5	1

Example 12.11. Our last example is a discrete version of the fractional continuous problem (Agrawal, 2008b, Example 2):

$$J_\alpha = \sum_{t=0}^{1} \frac{1}{2} \left({}_0\Delta_t^\alpha y(t) \right)^2 - y^\sigma(t) \longrightarrow \min, \quad y(0) = 0, \quad y(2) = 0. \quad (12.54)$$

The Euler–Lagrange extremal of (12.54) is easily obtained with the Maxima package (4 seconds):

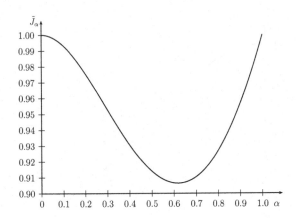

Fig. 12.3   Function $\tilde{J}_\alpha$ of Example 12.10 with $\gamma_1 = \gamma_2 = 1$, $b = 2$, $A = 0$, $B = 1$, and $\beta = \alpha$.

```
L3:(1/2)*v^2-u;$
extremal(L3,0,2,0,0,alpha,beta);
```

$$\tilde{y}(1) = \frac{1}{\alpha^2 + 1} \, . \tag{12.55}$$

For the particular case $\alpha = 1$ the equality (12.55) gives $\tilde{y}(1) = \frac{1}{2}$, which coincides with the solution to the non-fractional discrete problem

$$\sum_{t=0}^{1} \frac{1}{2} \left(\Delta y(t)\right)^2 - y^\sigma(t) = \sum_{t=0}^{1} \frac{1}{2} \left(y(t+1) - y(t)\right)^2 - y(t+1) \longrightarrow \min,$$

$$y(0) = 0 \, , \quad y(2) = 0 \, .$$

In Table 12.3 we show the values of $\tilde{y}(1)$ and $\tilde{J}_\alpha$ for some α's. As seen in Fig. 12.4, for $\alpha = 1$ one gets the maximum value of $\tilde{J}_\alpha$, $\alpha \in]0,1]$.

Table 12.3   Extremal values $\tilde{y}(1)$ of (12.54) for different α's

α	$\tilde{y}(1)$	$\tilde{J}_\alpha$
0.25	0.94117647058824	-0.47058823529412
0.50	0.8	-0.4
0.75	0.64	-0.32
1	0.5	-0.25

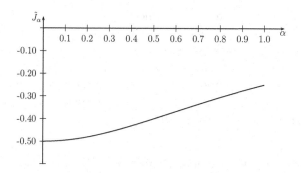

Fig. 12.4   Function $\tilde{J}_\alpha$ of Example 12.11.

12.5.6 *Conclusion*

In this section we introduce the study of fractional discrete-time problems of the calculus of variations of order α, $0 < \alpha \leq 1$, with left and right discrete operators of Riemann–Liouville type. For $\alpha = 1$ we obtain the classical discrete-time results of the calculus of variations (Kelley and Peterson, 1991).

Main results of the section include a fractional summation by parts formula (Theorem 12.8), a fractional discrete-time Euler–Lagrange equation (Theorem 12.9), transversality conditions (12.46) and (12.47), and a fractional discrete-time Legendre condition (Theorem 12.10). From the analysis of the results obtained from computer experiments, we conclude that when the value of α approaches one, the optimal value of the fractional discrete functional converges to the optimal value of the classical (non-fractional) discrete problem. On the other hand, the value of α for which the functional attains its minimum varies with the concrete problem under consideration.

The discrete-time calculus is also a very important tool in practical applications and in the modeling of real phenomena. Therefore, it is not a surprise that fractional discrete calculus is recently under strong development (see, e.g., (Baleanu and Jarad, 2006; Bastos, Ferreira and Torres, 2011b; Cresson, Frederico and Torres, 2009; Goodrich, 2010a,b) and references therein). There are some recent theses (Holm, 2010; Sengul, 2010) with results related with discrete calculus on time scale $\mathbb{T} = \{a, a + 1, a + 2, \ldots\}$. The thesis (Sengul, 2010) links the use of fractional difference equations with tumor growth modeling, and addresses a very important question: by changing the order of the difference equations from integers to fractional, in which conditions, are we able to provide more accurate models for real world problems? For more on the subject see (Pooseh, Rodrigues and Torres, 2011). Here we use fractional difference operators of Riemann–Liouville type. However, there are some results related with fractional difference operators in the Caputo sense (Anastassiou, 2011; Chen, Luo and Zhou, 2011), where the existence of solutions for IVP involving nonlinear fractional difference equations is discussed. The results of this section were first published in (Bastos, Ferreira and Torres, 2011a).

12.6   Fractional Variational Problems in $\mathbb{T} = (h\mathbb{Z})_a$

We now consider a discrete-time fractional calculus of variations on the time scale $(h\mathbb{Z})_a$, $h > 0$. First and second-order necessary optimality conditions are established. Examples illustrating the use of the new Euler–Lagrange and Legendre type conditions are given.

12.6.1   *Introduction*

Although the fractional Euler–Lagrange equations are obtained in a similar manner as in the standard variational calculus (Rabei *et al.*, 2007), some classical results are extremely difficult to be proved in a fractional context. This explains, for example, why a fractional Legendre type condition was absent from the literature of fractional variational calculus till very recently (Lazo and Torres, 2014). Here we introduce the concept of left and right fractional sum/difference (cf. Definition 12.21). In Section 12.6.2 we introduce notations, we give necessary definitions, and prove some preliminary results needed in the sequel. Main results appear in Section 12.6.3: we prove a fractional formula of h-summation by parts (Theorem 12.14), and necessary optimality conditions of first and second-order (Theorems 12.15 and 12.16, respectively) for the h-fractional problem of the calculus of variations (12.70). In Section 12.6.5 some illustrative examples are given, while some conclusions appear in Section 12.6.6.

12.6.2   *Preliminaries*

One way to approach the Riemann–Liouville fractional calculus is through the theory of linear differential equations (Podlubny, 1999). Miller and Ross (1989) use an analogous methodology to introduce fractional discrete operators for the case $\mathbb{T} = \mathbb{Z} = \{a, a+1, a+2, \ldots\}$, $a \in \mathbb{R}$. That was our starting point for the results presented in Section 12.5. Here we keep the graininess function constant but not, necessarily, equal to one: we use the theory of time scales in order to introduce fractional discrete operators for the more general case $\mathbb{T} = (h\mathbb{Z})_a = \{a, a + h, a + 2h, \ldots\}$, $a \in \mathbb{R}$, $h > 0$.

Before going further, we recall more results on the theory of time scales. Until we state the opposite, the following results are true for an arbitrary time scale.

For $n \in \mathbb{N}_0$ and rd-continuous functions $p_i : \mathbb{T} \to \mathbb{R}$, $1 \leq i \leq n$, let us consider the nth order linear dynamic equation

$$Ly = 0, \quad \text{where } Ly = y^{\Delta^n} + \sum_{i=1}^{n} p_i y^{\Delta^{n-i}}. \tag{12.56}$$

A function $y : \mathbb{T} \to \mathbb{R}$ is said to be a solution of equation (12.56) on $\mathbb{T}$ provided y is n times delta differentiable on $\mathbb{T}^{\kappa^n}$ and satisfies $Ly(t) = 0$ for all $t \in \mathbb{T}^{\kappa^n}$.

Lemma 12.6 (p. 239 of (Bohner and Peterson, 2001)).
If $z = (z_1, \ldots, z_n) : \mathbb{T} \to \mathbb{R}^n$ satisfies

$$z^{\Delta} = A(t)z(t), \quad \text{where} \quad A = \begin{pmatrix} 0 & 1 & 0 & \cdots & 0 \\ \vdots & 0 & 1 & \ddots & \vdots \\ \vdots & & \ddots & \ddots & 0 \\ 0 & \cdots\cdots & & 0 & 1 \\ -p_n & \cdots\cdots & & -p_2 & -p_1 \end{pmatrix} \tag{12.57}$$

for all $t \in \mathbb{T}^{\kappa}$, then $y = z_1$ is a solution of equation (12.56). Conversely, if y solves (12.56) on $\mathbb{T}$, then $z = \left(y, y^{\Delta}, \ldots, y^{\Delta^{n-1}} \right) : \mathbb{T} \to \mathbb{R}$ satisfies (12.57) for all $t \in \mathbb{T}^{\kappa^n}$.

Definition 12.18 (p. 239 of (Bohner and Peterson, 2001)). *We say that equation (12.56) is regressive provided $I + \mu(t)A(t)$ is invertible for all $t \in \mathbb{T}^{\kappa}$, where A is the matrix in (12.57).*

Definition 12.19 (p. 250 of (Bohner and Peterson, 2001)). *We define the Cauchy function $y : \mathbb{T} \times \mathbb{T}^{\kappa^n} \to \mathbb{R}$ for the linear dynamic equation (12.56) to be, for each fixed $s \in \mathbb{T}^{\kappa^n}$, the solution of the initial value problem*

$$Ly = 0, \quad y^{\Delta^i}(\sigma(s), s) = 0, \quad 0 \leq i \leq n - 2, \quad y^{\Delta^{n-1}}(\sigma(s), s) = 1. \tag{12.58}$$

Theorem 12.11 (p. 251 of (Bohner and Peterson, 2001)).
Suppose $\{y_1, \ldots, y_n\}$ is a fundamental system of the regressive equation
(12.56). Let $f \in C_{rd}$. Then the solution of the initial value problem

$$Ly = f(t), \quad y^{\Delta^i}(t_0) = 0, \quad 0 \leq i \leq n-1,$$

is given by $y(t) = \int_{t_0}^t y(t, s) f(s) \Delta s$, where $y(t, s)$ is the Cauchy function for
(12.56).

It is known that $y(t, s) := H_{n-1}(t, \sigma(s))$ is the Cauchy function for
$y^{\Delta^n} = 0$, where H_{n-1} is a time scale generalized polynomial (Bohner and
Peterson, 2001, Example 5.115). The generalized polynomials H_k are the
functions $H_k : \mathbb{T}^2 \to \mathbb{R}$, $k \in \mathbb{N}_0$, defined recursively as follows:

$$H_0(t, s) \equiv 1, \quad H_{k+1}(t, s) = \int_s^t H_k(\tau, s) \Delta \tau, \quad k = 1, 2, \ldots$$

for all $s, t \in \mathbb{T}$. If we let $H_k^{\Delta}(t, s)$ denote, for each fixed s, the derivative of
$H_k(t, s)$ with respect to t, then (cf. (Bohner and Peterson, 2001, p. 38))

$$H_k^{\Delta}(t, s) = H_{k-1}(t, s) \quad \text{for } k \in \mathbb{N}, \ t \in \mathbb{T}^{\kappa}.$$

Let $a \in \mathbb{R}$ and $h > 0$, $(h\mathbb{Z})_a = \{a, a+h, a+2h, \ldots\}$, and $b = a + kh$ for
some $k \in \mathbb{N}$.

From now on we restrict ourselves to the time scale $\mathbb{T} = (h\mathbb{Z})_a$, $h > 0$.
Our main goal is to propose and develop a discrete-time fractional variational theory in $\mathbb{T} = (h\mathbb{Z})_a$. We borrow the notations from the recent
calculus of variations on time scales (Bohner, 2004; Ferreira and Torres,
2008; Jarad, Baleanu and Maraaba, 2008). How to generalize the results
to an arbitrary time scale $\mathbb{T}$, with the graininess function μ depending on
time, is not clear and remains a challenging question.

We have $\sigma(t) = t+h$, $\rho(t) = t-h$, $\mu(t) \equiv h$, and we will frequently write
$f^{\sigma}(t) = f(\sigma(t))$. We put $\mathbb{T} = [a, b] \cap (h\mathbb{Z})_a$, so that $\mathbb{T}^{\kappa} = [a, \rho(b)] \cap (h\mathbb{Z})_a$
and $\mathbb{T}^{\kappa^2} = [a, \rho^2(b)] \cap (h\mathbb{Z})_a$. The delta derivative coincides in this case
with the forward h-difference:

$$f^{\Delta}(t) = \frac{f^{\sigma}(t) - f(t)}{\mu(t)}.$$

If $h = 1$, then we have the usual discrete forward difference $\Delta f(t)$. The delta
integral gives the h-sum (or h-integral) of f: $\int_a^b f(t) \Delta t = \sum_{k=\frac{a}{h}}^{\frac{b}{h}-1} f(kh)h$. If we
have a function f of two variables, $f(t, s)$, its partial forward h-differences
will be denoted by $\Delta_{t,h}$ and $\Delta_{s,h}$, respectively. We make use of the standard

conventions $\sum_{t=c}^{c-1} f(t) = 0$, $c \in \mathbb{Z}$, and $\prod_{i=0}^{-1} f(i) = 1$. Often, left fractional delta integration (resp., right fractional delta integration) of order $\nu > 0$ is denoted by $_a\Delta_t^{-\nu} f(t)$ (resp. $_t\Delta_b^{-\nu} f(t)$). Here, similarly as in Ross *et al.* (Ross, Samko and Love, 1994), where the authors omit the subscript t on the operator (the operator itself cannot depend on t), we write $_a\Delta_h^{-\nu} f(t)$ (resp. $_h\Delta_b^{-\nu} f(t)$).

Before giving an explicit formula for the generalized polynomials H_k on $h\mathbb{Z}$ we need the following definition, due to (Bastos, Ferreira and Torres, 2011b).

Definition 12.20. (Bastos, Ferreira and Torres, 2011b, Definition 2.6) *For arbitrary* $x, y \in \mathbb{R}$ *the h-factorial function is defined by*

$$x_h^{(y)} := h^y \frac{\Gamma(\frac{x}{h} + 1)}{\Gamma(\frac{x}{h} + 1 - y)},$$

where Γ *is the well-known Euler Gamma function, and we use the convention that division at a pole yields zero.*

Remark 12.16. Before proposing Definition 12.20 we have tried other possibilities. One that seemed most obvious is just to replace 1 by h in formula (12.24) because if we do $h = 1$ on the time scale $(h\mathbb{Z})_a$ we have the time scale $\mathbb{Z}$. This possibility did not reveal a good choice.

Remark 12.17. For $h = 1$, and in accordance with the previous similar definition on Section 12.5, we write $x^{(y)}$ to denote $x_h^{(y)}$.

Proposition 12.1. *For the time scale* $\mathbb{T} = (h\mathbb{Z})_a$ *one has*

$$H_k(t, s) := \frac{(t - s)_h^{(k)}}{k!} \quad \text{for all} \quad s, t \in \mathbb{T} \text{ and } k \in \mathbb{N}_0. \tag{12.59}$$

To prove (12.59) we use the following technical lemma. Throughout this chapter the basic property (12.5) of the Gamma function will be frequently used.

Lemma 12.7 (Lemma 2.7 of (Bastos, Ferreira and Torres, 2011b)). *Let* $s \in \mathbb{T}$. *Then, for all* $t \in \mathbb{T}^\kappa$ *one has*

$$\Delta_{t,h} \left\{ \frac{(t - s)_h^{(k+1)}}{(k+1)!} \right\} = \frac{(t - s)_h^{(k)}}{k!}.$$

Proof.   The equality follows by direct computations:

$$\Delta_{t,h} \left\{ \frac{(t-s)_h^{(k+1)}}{(k+1)!} \right\} = \frac{1}{h} \left\{ \frac{(\sigma(t)-s)_h^{(k+1)}}{(k+1)!} - \frac{(t-s)_h^{(k+1)}}{(k+1)!} \right\}$$

$$= \frac{h^{k+1}}{h(k+1)!} \left\{ \frac{\Gamma((t+h-s)/h+1)}{\Gamma((t+h-s)/h+1-(k+1))} - \frac{\Gamma((t-s)/h+1)}{\Gamma((t-s)/h+1-(k+1))} \right\}$$

$$= \frac{h^k}{(k+1)!} \left\{ \frac{((t-s)/h+1)\Gamma((t-s)/h+1)}{((t-s)/h-k)\Gamma((t-s)/h-k)} - \frac{\Gamma((t-s)/h+1)}{\Gamma((t-s)/h-k)} \right\}$$

$$= \frac{h^k}{k!} \left\{ \frac{\Gamma((t-s)/h+1)}{\Gamma((t-s)/h+1-k)} \right\} = \frac{(t-s)_h^{(k)}}{k!} . \qquad \square$$

Proof.   (of Proposition 12.1) We proceed by mathematical induction. For $k = 0$

$$H_0(t,s) = \frac{1}{0!} h^0 \frac{\Gamma(\frac{t-s}{h}+1)}{\Gamma(\frac{t-s}{h}+1-0)} = \frac{\Gamma(\frac{t-s}{h}+1)}{\Gamma(\frac{t-s}{h}+1)} = 1 .$$

Assume that (12.59) holds for k replaced by m. Then by Lemma 12.7

$$H_{m+1}(t,s) = \int_s^t H_m(\tau,s)\Delta\tau = \int_s^t \frac{(\tau-s)_h^{(m)}}{m!}\Delta\tau = \frac{(t-s)_h^{(m+1)}}{(m+1)!},$$

which is (12.59) with k replaced by $m+1$. $\qquad \square$

Let $y_1(t), \dots, y_n(t)$ be n linearly independent solutions of the linear homogeneous dynamic equation $y^{\Delta^n} = 0$. From Theorem 12.11 we know that the solution of (12.58) (with $L = \Delta^n$ and $t_0 = a$) is

$$y(t) = \Delta^{-n} f(t) = \int_a^t \frac{(t-\sigma(s))_h^{(n-1)}}{\Gamma(n)} f(s)\Delta s$$

$$= \frac{1}{\Gamma(n)} \sum_{k=a/h}^{t/h-1} (t-\sigma(kh))_h^{(n-1)} f(kh)h .$$

Since $y^{\Delta^i}(a) = 0$, $i = 0, \dots, n-1$, then we can write that

$$\Delta^{-n} f(t) = \frac{1}{\Gamma(n)} \sum_{k=a/h}^{t/h-n} (t-\sigma(kh))_h^{(n-1)} f(kh)h$$

$$= \frac{1}{\Gamma(n)} \int_a^{\sigma(t-nh)} (t-\sigma(s))_h^{(n-1)} f(s)\Delta s . \tag{12.60}$$

Note that function $t \to (\Delta^{-n} f)(t)$ is defined for $t = a + nh \mod(h)$ while function $t \to f(t)$ is defined for $t = a \mod(h)$. Extending (12.60) to any positive real value ν, and having as an analogy the continuous left and right fractional derivatives (Miller and Ross, 1993), we define the left fractional h-sum and the right fractional h-sum as introduced by (Bastos, Ferreira and Torres, 2011b). We denote by $\mathcal{F}_{\mathbb{T}}$ the set of all real valued functions defined on a given time scale $\mathbb{T}$.

Definition 12.21. (Bastos, Ferreira and Torres, 2011b, Definition 2.8) *Let* $a \in \mathbb{R}$, $h > 0$, $b = a + kh$ *with* $k \in \mathbb{N}$, *and put* $\mathbb{T} = [a,b] \cap (h\mathbb{Z})_a$. *Consider* $f \in \mathcal{F}_{\mathbb{T}}$. *The left and right fractional h-sum of order $\nu > 0$ are, respectively, the operators* ${}_a\Delta_h^{-\nu} : \mathcal{F}_{\mathbb{T}} \to \mathcal{F}_{\tilde{\mathbb{T}}_\nu^+}$ *and* ${}_h\Delta_b^{-\nu} : \mathcal{F}_{\mathbb{T}} \to \mathcal{F}_{\tilde{\mathbb{T}}_\nu^-}$, $\tilde{\mathbb{T}}_\nu^{\pm} = \{t \pm \nu h : t \in \mathbb{T}\}$, *defined by*

$$
{}_a\Delta_h^{-\nu} f(t) = \frac{1}{\Gamma(\nu)} \int_a^{\sigma(t-\nu h)} (t - \sigma(s))_h^{(\nu-1)} f(s) \Delta s
$$

$$
= \frac{1}{\Gamma(\nu)} \sum_{k=\frac{a}{h}}^{\frac{t}{h}-\nu} (t - \sigma(kh))_h^{(\nu-1)} f(kh) h,
$$

$$
{}_h\Delta_b^{-\nu} f(t) = \frac{1}{\Gamma(\nu)} \int_{t+\nu h}^{\sigma(b)} (s - \sigma(t))_h^{(\nu-1)} f(s) \Delta s
$$

$$
= \frac{1}{\Gamma(\nu)} \sum_{k=\frac{t}{h}+\nu}^{\frac{b}{h}} (kh - \sigma(t))_h^{(\nu-1)} f(kh) h.
$$

Remark 12.18. In Definition 12.21 we are using summations with limits that are reals. For example, the summation that appears in the definition of operator ${}_a\Delta_h^{-\nu}$ has the following meaning:

$$
\sum_{k=\frac{a}{h}}^{\frac{t}{h}-\nu} G(k) = G(a/h) + G(a/h+1) + G(a/h+2) + \cdots + G(t/h - \nu),
$$

where $t \in \{a + \nu h, a + h + \nu h, a + 2h + \nu h, \ldots, \underbrace{a + kh}_{b} + \nu h\}$ with $k \in \mathbb{N}$.

The next result was first proved in (Bastos, Ferreira and Torres, 2011b, Lemma 2.9).

Lemma 12.8. *Let $\nu > 0$ be an arbitrary positive real number. For any $t \in \mathbb{T}$ we have:*

(i) $\lim_{\nu \to 0} {}_a\Delta_h^{-\nu} f(t + \nu h) = f(t)$;
(ii) $\lim_{\nu \to 0} {}_h\Delta_b^{-\nu} f(t - \nu h) = f(t)$.

Proof.   Since

$$
\begin{aligned}
{}_a\Delta_h^{-\nu} f(t+\nu h) &= \frac{1}{\Gamma(\nu)} \int_a^{\sigma(t)} (t+\nu h - \sigma(s))_h^{(\nu-1)} f(s)\Delta s \\
&= \frac{1}{\Gamma(\nu)} \sum_{k=\frac{a}{h}}^{\frac{t}{h}} (t+\nu h - \sigma(kh))_h^{(\nu-1)} f(kh)h \\
&= h^\nu f(t) + \frac{\nu}{\Gamma(\nu+1)} \sum_{k=\frac{a}{h}}^{\frac{\rho(t)}{h}} (t+\nu h - \sigma(kh))_h^{(\nu-1)} f(kh)h ,
\end{aligned}
$$

it follows that $\lim_{\nu\to 0} {}_a\Delta_h^{-\nu} f(t+\nu h) = f(t)$. The proof of (ii) is similar. $\square$

For any $t \in \mathbb{T}$ and for any $\nu \geq 0$ we define ${}_a\Delta_h^0 f(t) := {}_h\Delta_b^0 f(t) := f(t)$ and write

$$
\begin{aligned}
{}_a\Delta_h^{-\nu} f(t+\nu h) &= h^\nu f(t) + \frac{\nu}{\Gamma(\nu+1)} \int_a^t (t+\nu h - \sigma(s))_h^{(\nu-1)} f(s)\Delta s , \\
{}_h\Delta_b^{-\nu} f(t) &= h^\nu f(t-\nu h) + \frac{\nu}{\Gamma(\nu+1)} \int_{\sigma(t)}^{\sigma(b)} (s+\nu h - \sigma(t))_h^{(\nu-1)} f(s)\Delta s .
\end{aligned}
$$

$$(12.61)$$

Theorem 12.12. *Let $f \in \mathcal{F}_\mathbb{T}$ and $\nu \geq 0$. For all $t \in \mathbb{T}^\kappa$ we have*

$$
{}_a\Delta_h^{-\nu} f^\Delta(t+\nu h) = ({}_a\Delta_h^{-\nu} f(t+\nu h))^\Delta - \frac{\nu}{\Gamma(\nu+1)}(t+\nu h - a)_h^{(\nu-1)} f(a) .
$$

$$(12.62)$$

To prove Theorem 12.12 we make use of a technical lemma:

Lemma 12.9. *Let $t \in \mathbb{T}^\kappa$. The following equality holds for all $s \in \mathbb{T}^\kappa$:*

$$
\Delta_{s,h}\left((t+\nu h - s)_h^{(\nu-1)} f(s) \right)
$$
$$
= (t+\nu h - \sigma(s))_h^{(\nu-1)} f^\Delta(s) - (v-1)(t+\nu h - \sigma(s))_h^{(\nu-2)} f(s) . \quad (12.63)
$$

Proof. Direct calculations give the intended result:

$$\Delta_{s,h}\left((t + \nu h - s)_h^{(\nu-1)} f(s)\right)$$

$$= \Delta_{s,h}\left((t + \nu h - s)_h^{(\nu-1)}\right) f(s) + (t + \nu h - \sigma(s))_h^{(\nu-1)} f^\Delta(s)$$

$$= \frac{f(s)}{h}\left[\frac{h^{\nu-1}\Gamma\left(\frac{t+\nu h-\sigma(s)}{h}+1\right)}{\Gamma\left(\frac{t+\nu h-\sigma(s)}{h}+1-(\nu-1)\right)} - \frac{h^{\nu-1}\Gamma\left(\frac{t+\nu h-s}{h}+1\right)}{\Gamma\left(\frac{t+\nu h-s}{h}+1-(\nu-1)\right)}\right]$$

$$+ (t + \nu h - \sigma(s))_h^{(\nu-1)} f^\Delta(s)$$

$$= f(s)h^{\nu-2}\left[\frac{\Gamma(\frac{t+\nu h-s}{h})}{\Gamma(\frac{t-s}{h}+1)} - \frac{\Gamma(\frac{t+\nu h-s}{h}+1)}{\Gamma(\frac{t-s}{h}+2)}\right] + (t + \nu h - \sigma(s))_h^{(\nu-1)} f^\Delta(s)$$

$$= \frac{f(s)h^{\nu-2}\Gamma(\frac{t+\nu h-s-h}{h}+1)}{\Gamma(\frac{t-s+\nu h-h}{h}+1-(\nu-2))}(-(\nu-1)) + (t + \nu h - \sigma(s))_h^{(\nu-1)} f^\Delta(s)$$

$$= -(\nu-1)(t + \nu h - \sigma(s))_h^{(\nu-2)} f(s) + (t + \nu h - \sigma(s))_h^{(\nu-1)} f^\Delta(s),$$

where the first equality follows directly from (12.11). $\qquad\square$

Remark 12.19. Given an arbitrary $t \in \mathbb{T}^\kappa$ it is easy to prove, in a similar way as in the proof of Lemma 12.9, the following equality analogous to (12.63): for all $s \in \mathbb{T}^\kappa$

$$\Delta_{s,h}\left((s + \nu h - \sigma(t))_h^{(\nu-1)} f(s)\right)$$

$$= (\nu-1)(s + \nu h - \sigma(t))_h^{(\nu-2)} f^\sigma(s) + (s + \nu h - \sigma(t))_h^{(\nu-1)} f^\Delta(s). \quad (12.64)$$

Proof. (of Theorem 12.12) From Lemma 12.9 we obtain that

$${}_a\Delta_h^{-\nu} f^\Delta(t + \nu h) = h^\nu f^\Delta(t) + \frac{\nu}{\Gamma(\nu+1)}\int_a^t (t + \nu h - \sigma(s))_h^{(\nu-1)} f^\Delta(s)\Delta s$$

$$= h^\nu f^\Delta(t) + \frac{\nu}{\Gamma(\nu+1)}\left[(t + \nu h - s)_h^{(\nu-1)} f(s)\right]_{s=a}^{s=t}$$

$$+ \frac{\nu}{\Gamma(\nu+1)}\int_a^{\sigma(t)} (\nu-1)(t + \nu h - \sigma(s))_h^{(\nu-2)} f(s)\Delta s$$

$$= -\frac{\nu(t + \nu h - a)_h^{(\nu-1)}}{\Gamma(\nu+1)} f(a) + h^\nu f^\Delta(t) + \nu h^{\nu-1} f(t)$$

$$+ \frac{\nu}{\Gamma(\nu+1)}\int_a^t (\nu-1)(t + \nu h - \sigma(s))_h^{(\nu-2)} f(s)\Delta s.$$

$$(12.65)$$

We now show that $({}_a\Delta_h^{-\nu} f(t+\nu h))^\Delta$ equals (12.65):

$$({}_a\Delta_h^{-\nu} f(t+\nu h))^\Delta$$

$$= \frac{1}{h}\left[h^\nu f(\sigma(t)) + \frac{\nu}{\Gamma(\nu+1)} \int_a^{\sigma(t)} (\sigma(t)+\nu h - \sigma(s))_h^{(\nu-1)} f(s)\Delta s \right.$$

$$\left. - h^\nu f(t) - \frac{\nu}{\Gamma(\nu+1)} \int_a^t (t+\nu h - \sigma(s))_h^{(\nu-1)} f(s)\Delta s \right]$$

$$= h^\nu f^\Delta(t) + \frac{\nu}{h\Gamma(\nu+1)}\left[\int_a^t (\sigma(t)+\nu h - \sigma(s))_h^{(\nu-1)} f(s)\Delta s \right.$$

$$\left. - \int_a^t (t+\nu h - \sigma(s))_h^{(\nu-1)} f(s)\Delta s\right] + h^{\nu-1}\nu f(t)$$

$$= h^\nu f^\Delta(t) + \frac{\nu}{\Gamma(\nu+1)}\int_a^t \Delta_{t,h}\left((t+\nu h - \sigma(s))_h^{(\nu-1)}\right) f(s)\Delta s$$

$$+ h^{\nu-1}\nu f(t)$$

$$= h^\nu f^\Delta(t) + \frac{\nu}{\Gamma(\nu+1)}\int_a^t (\nu-1)(t+\nu h - \sigma(s))_h^{(\nu-2)} f(s)\Delta s$$

$$+ \nu h^{\nu-1} f(t).$$

$\square$

Follows the counterpart of Theorem 12.12 for the right fractional h-sum:

Theorem 12.13. *Let $f \in \mathcal{F}_\mathbb{T}$ and $\nu \geq 0$. For all $t \in \mathbb{T}^\kappa$ we have*

$$_h\Delta_{\rho(b)}^{-\nu} f^\Delta(t-\nu h) = \frac{\nu}{\Gamma(\nu+1)}(b+\nu h - \sigma(t))_h^{(\nu-1)} f(b) + ({}_h\Delta_b^{-\nu} f(t-\nu h))^\Delta.$$

$$(12.66)$$

Proof. From (12.64) we obtain from integration by parts (item 2 of Lemma 12.1) that

$$_h\Delta_{\rho(b)}^{-\nu} f^\Delta(t-\nu h) = \frac{\nu(b+\nu h - \sigma(t))_h^{(\nu-1)}}{\Gamma(\nu+1)} f(b) + h^\nu f^\Delta(t) - \nu h^{\nu-1} f(\sigma(t))$$

$$- \frac{\nu}{\Gamma(\nu+1)}\int_{\sigma(t)}^b (\nu-1)(s+\nu h - \sigma(t))_h^{(\nu-2)} f^\sigma(s)\Delta s. \quad (12.67)$$

We show that $(_h\Delta_b^{-\nu}f(t-\nu h))^\Delta$ equals (12.67):

$$(_h\Delta_b^{-\nu}f(t-\nu h))^\Delta$$

$$= h^\nu f^\Delta(t) + \frac{\nu}{h\Gamma(\nu+1)}\left[\int_{\sigma^2(t)}^{\sigma(b)}(s+\nu h-\sigma^2(t))_h^{(\nu-1)}f(s)\Delta s\right.$$

$$\left.- \int_{\sigma^2(t)}^{\sigma(b)}(s+\nu h-\sigma(t))_h^{(\nu-1)}f(s)\Delta s\right] - \nu h^{\nu-1}f(\sigma(t))$$

$$= h^\nu f^\Delta(t) + \frac{\nu}{\Gamma(\nu+1)}\int_{\sigma^2(t)}^{\sigma(b)}\Delta_{t,h}\left((s+\nu h-\sigma(t))_h^{(\nu-1)}\right)f(s)\Delta s$$

$$- \nu h^{\nu-1}f(\sigma(t))$$

$$= h^\nu f^\Delta(t) - \frac{\nu}{\Gamma(\nu+1)}\int_{\sigma^2(t)}^{\sigma(b)}(\nu-1)(s+\nu h-\sigma^2(t))_h^{(\nu-2)}f(s)\Delta s$$

$$- \nu h^{\nu-1}f(\sigma(t))$$

$$= h^\nu f^\Delta(t) - \frac{\nu}{\Gamma(\nu+1)}\int_{\sigma(t)}^{b}(\nu-1)(s+\nu h-\sigma(t))_h^{(\nu-2)}f(s)\Delta s$$

$$- \nu h^{\nu-1}f(\sigma(t)). \qquad \square$$

Definition 12.22. *Let $0 < \alpha \leq 1$ and set $\gamma := 1 - \alpha$. The left fractional difference $_a\Delta_h^\alpha f(t)$ and the right fractional difference $_h\Delta_b^\alpha f(t)$ of order α of a function $f \in \mathcal{F}_\mathbb{T}$ are defined as*

$$_a\Delta_h^\alpha f(t) := (_a\Delta_h^{-\gamma}f(t+\gamma h))^\Delta \quad \text{and} \quad _h\Delta_b^\alpha f(t) := -(_h\Delta_b^{-\gamma}f(t-\gamma h))^\Delta$$

for all $t \in \mathbb{T}^\kappa$.

12.6.3 *Fractional h-summation by parts*

Our aim is to introduce the h-fractional problem of the calculus of variations and to prove corresponding necessary optimality conditions. In order to obtain an Euler–Lagrange type equation (cf. Theorem 12.15) we first prove a fractional formula of h-summation by parts. For that we make use of the following lemma.

Lemma 12.10. *Let f and k be two functions defined on $\mathbb{T}^\kappa$ and $\mathbb{T}^{\kappa^2}$, respectively, and g a function defined on $\mathbb{T}^\kappa \times \mathbb{T}^{\kappa^2}$. The following equality holds:*

$$\int_a^b f(t)\left[\int_a^t g(t,s)k(s)\Delta s\right]\Delta t = \int_a^{\rho(b)}k(t)\left[\int_{\sigma(t)}^b g(s,t)f(s)\Delta s\right]\Delta t.$$

Proof. Consider the matrices

$$R = [f(a+h), f(a+2h), \cdots, f(b-h)],$$

$$C_1 = \begin{bmatrix} g(a+h,a)k(a) \\ g(a+2h,a)k(a) + g(a+2h,a+h)k(a+h) \\ \vdots \\ g(b-h,a)k(a) + g(b-h,a+h)k(a+h) + \cdots + g(b-h,b-2h)k(b-2h) \end{bmatrix},$$

$$C_2 = \begin{bmatrix} g(a+h,a) \\ g(a+2h,a) \\ \vdots \\ g(b-h,a) \end{bmatrix},$$

$$C_3 = \begin{bmatrix} 0 \\ g(a+2h,a+h) \\ \vdots \\ g(b-h,a+h) \end{bmatrix},$$

$$\vdots$$

$$C_k = \begin{bmatrix} 0 \\ 0 \\ \vdots \\ g(b-h,b-2h) \end{bmatrix}.$$

Direct calculations show that

$$\int_a^b f(t) \left[\int_a^t g(t,s)k(s)\Delta s \right] \Delta t$$

$$= h^2 \sum_{i=a/h}^{b/h-1} f(ih) \sum_{j=a/h}^{i-1} g(ih,jh)k(jh) = h^2 R \cdot C_1$$

$$= h^2 R \cdot [k(a)C_2 + k(a+h)C_3 + \cdots + k(b-2h)C_k]$$

$$= h^2 \left[k(a) \sum_{j=a/h+1}^{b/h-1} g(jh,a)f(jh) + k(a+h) \sum_{j=a/h+2}^{b/h-1} g(jh,a+h)f(jh) \right.$$

$$\left. + \cdots + k(b-2h) \sum_{j=b/h-1}^{b/h-1} g(jh,b-2h)f(jh) \right]$$

$$= \sum_{i=a/h}^{b/h-2} k(ih)h \sum_{j=\sigma(ih)/h}^{b/h-1} g(jh,ih)f(jh)h$$

$$= \int_a^{\rho(b)} k(t) \left[\int_{\sigma(t)}^b g(s,t)f(s)\Delta s \right] \Delta t.$$

□

The following fractional h-summation by parts was first given in (Bastos, Ferreira and Torres, 2011b, Theorem 3.2).

Theorem 12.14 (Fractional h-summation by parts). *Let f and g be real valued functions defined on $\mathbb{T}^\kappa$ and $\mathbb{T}$, respectively. Fix $0 < \alpha \leq 1$ and put $\gamma := 1 - \alpha$. Then,*

$$\int_a^b f(t)_a\Delta_h^\alpha g(t)\Delta t$$

$$= h^\gamma f(\rho(b))g(b) - h^\gamma f(a)g(a) + \int_a^{\rho(b)} {}_h\Delta_{\rho(b)}^\alpha f(t)g^\sigma(t)\Delta t + \frac{\gamma g(a)}{\Gamma(\gamma+1)}$$

$$\times \left(\int_a^b (t+\gamma h - a)_h^{(\gamma-1)} f(t)\Delta t - \int_{\sigma(a)}^b (t+\gamma h - \sigma(a))_h^{(\gamma-1)} f(t)\Delta t \right).$$

(12.68)

Proof. By (12.02) we can write

$$\int_a^b f(t)_a\Delta_h^\alpha g(t)\Delta t = \int_a^b f(t)(_a\Delta_h^{-\gamma}g(t+\gamma h))^\Delta \Delta t$$

$$= \int_a^b f(t) \left[{}_a\Delta_h^{-\gamma}g^\Delta(t+\gamma h) + \frac{\gamma}{\Gamma(\gamma+1)}(t+\gamma h - a)_h^{(\gamma-1)}g(a) \right] \Delta t$$

$$= \int_a^b f(t)_a\Delta_h^{-\gamma}g^\Delta(t+\gamma h)\Delta t + \int_a^b \frac{\gamma}{\Gamma(\gamma+1)}(t+\gamma h - a)_h^{(\gamma-1)} f(t)g(a)\Delta t.$$

(12.69)

Using (12.61) we get

$$\int_a^b f(t)\,_a\Delta_h^{-\gamma}g^\Delta(t+\gamma h)\Delta t$$

$$= \int_a^b f(t)\left[h^\gamma g^\Delta(t) + \frac{\gamma}{\Gamma(\gamma+1)}\int_a^t (t+\gamma h - \sigma(s))_h^{(\gamma-1)}g^\Delta(s)\Delta s\right]\Delta t$$

$$= h^\gamma \int_a^b f(t)g^\Delta(t)\Delta t$$

$$+ \frac{\gamma}{\Gamma(\gamma+1)}\int_a^{\rho(b)} g^\Delta(t)\int_{\sigma(t)}^b (s+\gamma h - \sigma(t))_h^{(\gamma-1)}f(s)\Delta s\Delta t$$

$$= h^\gamma f(\rho(b))[g(b) - g(\rho(b))] + \int_a^{\rho(b)} g^\Delta(t)\,_h\Delta_{\rho(b)}^{-\gamma}f(t-\gamma h)\Delta t,$$

where the third equality follows by Lemma 12.10. We proceed to develop the right-hand side of the last equality as follows:

$$h^\gamma f(\rho(b))[g(b) - g(\rho(b))] + \int_a^{\rho(b)} g^\Delta(t)\,_h\Delta_{\rho(b)}^{-\gamma}f(t-\gamma h)\Delta t$$

$$= h^\gamma f(\rho(b))[g(b) - g(\rho(b))] + \left[g(t)\,_h\Delta_{\rho(b)}^{-\gamma}f(t-\gamma h)\right]_{t=a}^{t=\rho(b)}$$

$$- \int_a^{\rho(b)} g^\sigma(t)(\,_h\Delta_{\rho(b)}^{-\gamma}f(t-\gamma h))^\Delta\Delta t$$

$$= h^\gamma f(\rho(b))g(b) - h^\gamma f(a)g(a)$$

$$- \frac{\gamma}{\Gamma(\gamma+1)}g(a)\int_{\sigma(a)}^b (s+\gamma h - \sigma(a))_h^{(\gamma-1)}f(s)\Delta s$$

$$+ \int_a^{\rho(b)} \left(\,_h\Delta_{\rho(b)}^\alpha f(t)\right)g^\sigma(t)\Delta t,$$

where the first equality follows from Lemma 12.1. Putting this into (12.69) we get (12.68). $\qquad\qquad\square$

12.6.4 *Necessary optimality conditions*

We begin to fix two arbitrary real numbers α and β such that $\alpha, \beta \in (0,1]$. Further, we put $\gamma := 1 - \alpha$ and $\nu := 1 - \beta$.

Let a function $L(t, u, v, w) : \mathbb{T}^\kappa \times \mathbb{R} \times \mathbb{R} \times \mathbb{R} \to \mathbb{R}$ be given. We consider the problem of minimizing (or maximizing) a functional $\mathcal{L} : \mathcal{F}_\mathbb{T} \to \mathbb{R}$ subject to given boundary conditions:

$$\mathcal{L}[y] = \int_a^b L(t, y^\sigma(t), {}_a\Delta_h^\alpha y(t), {}_h\Delta_b^\beta y(t))\Delta t \longrightarrow \min,$$

$$y(a) = A, \ y(b) = B.$$

(12.70)

Our main aim is to derive necessary optimality conditions for problem (12.70).

Definition 12.23. *For* $f \in \mathcal{F}_\mathbb{T}$ *we define the norm*

$$\|f\| = \max_{t \in \mathbb{T}^\kappa} |f^\sigma(t)| + \max_{t \in \mathbb{T}^\kappa} |{}_a\Delta_h^\alpha f(t)| + \max_{t \in \mathbb{T}^\kappa} |{}_h\Delta_b^\beta f(t)|.$$

A function $\hat{y} \in \mathcal{F}_\mathbb{T}$ *with* $\hat{y}(a) = A$ *and* $\hat{y}(b) = B$ *is called a local minimizer for problem* (12.70) *provided there exists* $\delta > 0$ *such that* $\mathcal{L}[\hat{y}] \leq \mathcal{L}[y]$ *for all* $y \in \mathcal{F}_\mathbb{T}$ *with* $y(a) = A$ *and* $y(b) = B$ *and* $\|y - \hat{y}\| < \delta$.

Definition 12.24. *A function* $\eta \in \mathcal{F}_\mathbb{T}$ *is called an admissible variation provided* $\eta \neq 0$ *and* $\eta(a) = \eta(b) = 0$.

From now on we assume that the second-order partial derivatives L_{uu}, L_{uv}, L_{uw}, L_{vw}, L_{vv}, and L_{ww} exist and are continuous.

12.6.4.1 *First-order optimality condition*

The next theorem gives a first-order necessary condition for problem (12.70), i.e., an Euler–Lagrange type equation for the fractional h-difference setting. It was first given in (Bastos, Ferreira and Torres, 2011b, Theorem 3.5).

Theorem 12.15. *If* $\hat{y} \in \mathcal{F}_\mathbb{T}$ *is a local minimizer for problem* (12.70), *then the equality*

$$L_u[\hat{y}](t) + {}_h\Delta_{\rho(b)}^\alpha L_v[\hat{y}](t) + {}_a\Delta_h^\beta L_w[\hat{y}](t) = 0 \qquad (12.71)$$

holds for all $t \in \mathbb{T}^{\kappa^2}$ *with operator* $[\cdot]$ *defined by*

$$[y](s) = \left(s, y^\sigma(s), {}_a\Delta_s^\alpha y(s), {}_s\Delta_b^\beta y(s)\right).$$

Proof. Suppose that $\hat{y}$ is a local minimizer of $\mathcal{L}$. Let η be an arbitrarily fixed admissible variation and define a function

$$\Phi : \left(-\frac{\delta}{\|\eta\|}, \frac{\delta}{\|\eta\|}\right) \to \mathbb{R}$$

by

$$\Phi(\varepsilon) = \mathcal{L}[\hat{y} + \varepsilon\eta]. \qquad (12.72)$$

This function has a minimum at $\varepsilon = 0$, so we must have $\Phi'(0) = 0$, i.e.,

$$\int_a^b \left[L_u[\hat{y}](t)\eta^\sigma(t) + L_v[\hat{y}](t)\,_a\Delta_h^\alpha \eta(t) + L_w[\hat{y}](t)\,_h\Delta_b^\beta \eta(t) \right] \Delta t = 0,$$

which we may write equivalently as

$$hL_u[\hat{y}](t)\eta^\sigma(t)|_{t=\rho(b)} + \int_a^{\rho(b)} L_u[\hat{y}](t)\eta^\sigma(t)\Delta t + \int_a^b L_v[\hat{y}](t)\,_a\Delta_h^\alpha \eta(t)\Delta t$$

$$+ \int_a^b L_w[\hat{y}](t)\,_h\Delta_b^\beta \eta(t)\Delta t = 0. \quad (12.73)$$

Using Theorem 12.14 and the fact that $\eta(a) = \eta(b) = 0$, we get

$$\int_a^b L_v[\hat{y}](t)\,_a\Delta_h^\alpha \eta(t)\Delta t = \int_a^{\rho(b)} \left({}_h\Delta_{\rho(b)}^\alpha \left(L_v[\hat{y}] \right)(t) \right) \eta^\sigma(t)\Delta t \quad (12.74)$$

for the third term in (12.73). Using (12.66) it follows that

$$\int_a^b L_w[\hat{y}](t)\,_h\Delta_b^\beta \eta(t)\Delta t$$

$$= -\int_a^h L_w[\hat{y}](t)(\,_h\Delta_b^{-\nu}\eta(t-\nu h))^\Delta \Delta t$$

$$= -\int_a^b L_w[\hat{y}](t) \left[{}_h\Delta_{\rho(b)}^{-\nu}\eta^\Delta(t-\nu h) - \frac{\nu(b+\nu h - \sigma(t))_h^{(\nu-1)}\eta(b)}{\Gamma(\nu+1)} \right] \Delta t$$

$$= -\int_a^b L_w[\hat{y}](t)\,_h\Delta_{\rho(b)}^{-\nu}\eta^\Delta(t-\nu h)\Delta t$$

$$+ \frac{\nu\eta(b)}{\Gamma(\nu+1)} \int_a^b (b+\nu h - \sigma(t))_h^{(\nu-1)} L_w[\hat{y}](t)\Delta t.$$

$$(12.75)$$

We now use Lemma 12.10 to get

$$\int_a^b L_w[\hat{y}](t)\,{}_h\Delta_{\rho(b)}^{-\nu}\eta^\Delta(t-\nu h)\Delta t$$

$$= \int_a^b L_w[\hat{y}](t)\left[h^\nu\eta^\Delta(t)\right.$$

$$\left. + \frac{\nu}{\Gamma(\nu+1)}\int_{\sigma(t)}^b (s+\nu h-\sigma(t))_h^{(\nu-1)}\eta^\Delta(s)\Delta s\right]\Delta t$$

$$= \int_a^b h^\nu L_w[\hat{y}](t)\eta^\Delta(t)\Delta t$$

$$+ \frac{\nu}{\Gamma(\nu+1)}\int_a^{\rho(b)}\left[L_w[\hat{y}](t)\int_{\sigma(t)}^b (s+\nu h-\sigma(t))_h^{(\nu-1)}\eta^\Delta(s)\Delta s\right]\Delta t$$

$$= \int_a^b h^\nu L_w[\hat{y}](t)\eta^\Delta(t)\Delta t$$

$$+ \frac{\nu}{\Gamma(\nu+1)}\int_a^b\left[\eta^\Delta(t)\int_a^t (t+\nu h-\sigma(s))_h^{(\nu-1)}L_w[\hat{y}](s)\Delta s\right]\Delta t$$

$$= \int_a^b \eta^\Delta(t)\,{}_a\Delta_h^{-\nu}\left(L_w[\hat{y}]\right)(t+\nu h)\Delta t.$$

$$(12.76)$$

We apply again the time scale integration by parts formula (Lemma 12.1), this time to (12.76), to obtain,

$$\int_a^b \eta^\Delta(t)\,{}_a\Delta_h^{-\nu}\left(L_w[\hat{y}]\right)(t+\nu h)\Delta t$$

$$= \int_a^{\rho(b)} \eta^\Delta(t)\,{}_a\Delta_h^{-\nu}\left(L_w[\hat{y}]\right)(t+\nu h)\Delta t$$

$$+ (\eta(b)-\eta(\rho(b)))\,{}_a\Delta_h^{-\nu}\left(L_w[\hat{y}]\right)(t+\nu h)|_{t=\rho(b)}$$

$$= \left[\eta(t)\,{}_a\Delta_h^{-\nu}\left(L_w[\hat{y}]\right)(t+\nu h)\right]_{t=a}^{t=\rho(b)}$$

$$- \int_a^{\rho(b)} \eta^\sigma(t)(\,{}_a\Delta_h^{-\nu}\left(L_w[\hat{y}]\right)(t+\nu h))^\Delta\Delta t$$

$$+ \eta(b)\,{}_a\Delta_h^{-\nu}\left(L_w[\hat{y}]\right)(t+\nu h)|_{t=\rho(b)}$$

$$- \eta(\rho(b))\,{}_a\Delta_h^{-\nu}\left(L_w[\hat{y}]\right)(t+\nu h)|_{t=\rho(b)}$$

$$= \eta(b)\,{}_a\Delta_h^{-\nu}\left(L_w[\hat{y}]\right)(t+\nu h)|_{t=\rho(b)} - \eta(a)\,{}_a\Delta_h^{-\nu}\left(L_w[\hat{y}]\right)(t+\nu h)|_{t=a}$$

$$- \int_a^{\rho(b)} \eta^\sigma(t)\,{}_a\Delta_h^\beta\left(L_w[\hat{y}]\right)(t)\Delta t.$$

$$(12.77)$$

Since $\eta(a) = \eta(b) = 0$ we obtain, from (12.76) and (12.77), that

$$\int_a^b L_w[\hat{y}](t)\,_h\Delta_{\rho(b)}^{-\nu}\eta^\Delta(t)\Delta t = -\int_a^{\rho(b)} \eta^\sigma(t)\,_a\Delta_h^\beta (L_w[\hat{y}])(t)\Delta t\,,$$

and after inserting in (12.75), that

$$\int_a^b L_w[\hat{y}](t)\,_h\Delta_b^\beta\eta(t)\Delta t = \int_a^{\rho(b)} \eta^\sigma(t)\,_a\Delta_h^\beta (L_w[\hat{y}])(t)\Delta t. \tag{12.78}$$

By (12.74) and (12.78) we may write (12.73) as

$$\int_a^{\rho(b)} \left[L_u[\hat{y}](t) + \,_h\Delta_{\rho(b)}^\alpha (L_v[\hat{y}])(t) + \,_a\Delta_h^\beta (L_w[\hat{y}])(t) \right] \eta^\sigma(t)\Delta t = 0\,.$$

Since the values of $\eta^\sigma(t)$ are arbitrary for $t \in \mathbb{T}^{\kappa^2}$, the Euler–Lagrange equation (12.71) holds along $\hat{y}$. □

The next result is a direct corollary of Theorem 12.15 (cf., e.g., (Bohner, 2004; Ferreira and Torres, 2008)).

Corollary 12.3 (The h-Euler–Lagrange equation). *Let $\mathbb{T}$ be the time scale $h\mathbb{Z}$, $h > 0$, with the forward jump operator σ and the delta derivative Δ. Assume $a, b \in \mathbb{T}$, $a < b$. If $\hat{y}$ is a solution to the problem*

$$\mathcal{L}[y] = \int_a^b L(t, y^\sigma(t), y^\Delta(t))\Delta t \longrightarrow \min, \ y(a) = A, \ y(b) = B\,,$$

then the equality $L_u(t, \hat{y}^\sigma(t), \hat{y}^\Delta(t)) - \left(L_v(t, \hat{y}^\sigma(t), \hat{y}^\Delta(t))\right)^\Delta = 0$ holds for all $t \in \mathbb{T}^{\kappa^2}$.

Proof. Choose $\alpha = 1$ and a L that does not depend on w in Theorem 12.15. □

Remark 12.20. If we take $h = 1$ in Corollary 12.3 we have that

$$L_u(t, \hat{y}^\sigma(t), \Delta\hat{y}(t)) - \Delta L_v(t, \hat{y}^\sigma(t), \Delta\hat{y}(t)) = 0$$

holds for all $t \in \mathbb{T}^{\kappa^2}$. This equation is usually called the discrete Euler–Lagrange equation, and can be found, e.g., in (Kelley and Peterson, 1991, Chap. 8).

12.6.4.2 *Natural boundary conditions*

If the initial condition $y(a) = A$ is not present in problem (12.70) (i.e., $y(a)$ is free), besides the h-fractional Euler–Lagrange equation (12.71) the following supplementary condition must be fulfilled:

$$-h^\gamma L_v[\hat{y}](a) + \frac{\gamma}{\Gamma(\gamma+1)} \left(\int_a^b (t + \gamma h - a)_h^{(\gamma-1)} L_v[\hat{y}](t)\Delta t \right.$$

$$\left. - \int_{\sigma(a)}^b (t + \gamma h - \sigma(a))_h^{(\gamma-1)} L_v[\hat{y}](t)\Delta t \right) + L_w[\hat{y}](a) = 0. \tag{12.79}$$

Similarly, if $y(b) = B$ is not present in (12.70) ($y(b)$ is free), the extra condition

$$hL_u[\hat{y}](\rho(b)) + h^\gamma L_v[\hat{y}](\rho(b)) - h^\nu L_w[\hat{y}](\rho(b))$$

$$+ \frac{\nu}{\Gamma(\nu+1)} \left(\int_a^b (b + \nu h - \sigma(t))_h^{(\nu-1)} L_w[\hat{y}](t)\Delta t \right.$$

$$\left. - \int_a^{\rho(b)} (\rho(b) + \nu h - \sigma(t))_h^{(\nu-1)} L_w[\hat{y}](t)\Delta t \right) = 0 \tag{12.80}$$

is added to Theorem 12.15. We leave the proof of the natural boundary conditions (12.79) and (12.80) to the reader. We just note here that the first term in (12.80) arises from the first term of the left-hand side of (12.73).

12.6.4.3 *Second-order optimality condition*

We now obtain a second-order necessary condition for problem (12.70), i.e., we prove a Legendre optimality type condition for the fractional h-difference setting. The result is due to (Bastos, Ferreira and Torres, 2011b).

Theorem 12.16 (The h-fractional Legendre necessary condition).
If $\hat{y} \in \mathcal{F}_\mathbb{T}$ is a local minimizer for problem (12.70), then the inequality

$$h^2 L_{uu}[\hat{y}](t) + 2h \left(h^\gamma L_{uv}[\hat{y}](t) + h^\nu(\nu-1)L_{uw}[\hat{y}](t) \right)$$

$$+ 2h^{\nu+\gamma} \left[(\gamma-1)L_{vw}[\hat{y}](\sigma(t)) + (\nu-1)L_{vw}[\hat{y}](t) \right] + \left(h^\nu(\nu-1) \right)^2 L_{ww}[\hat{y}](t)$$

$$+ h^{2\nu} L_{ww}[\hat{y}](\sigma(t)) + \int_a^t h^3 L_{ww}[\hat{y}](s) \left(\frac{\nu(1-\nu)}{\Gamma(\nu+1)} (t + \nu h - \sigma(s))_h^{(\nu-2)} \right)^2 \Delta s$$

$$+ h^\gamma L_{vv}[\hat{y}](t) + \left(h^\gamma(\gamma-1) \right)^2 L_{vv}[\hat{y}](\sigma(t))$$

$$+ \int_{\sigma(\sigma(t))}^b h^3 L_{vv}[\hat{y}](s) \left(\frac{\gamma(\gamma-1)(s + \gamma h - \sigma(\sigma(t)))_h^{(\gamma-2)}}{\Gamma(\gamma+1)} \right)^2 \Delta s \geq 0 \tag{12.81}$$

holds for all $t \in \mathbb{T}^{\kappa^2}$, *where* $[\hat{y}](t) = \left(t, \hat{y}^\sigma(t), {}_a\Delta_t^\alpha \hat{y}(t), {}_t\Delta_b^\beta \hat{y}(t)\right)$.

Proof.   By the hypothesis of the theorem, and letting Φ be as in (12.72), we have a necessary optimality condition that $\Phi''(0) \geq 0$ for an arbitrary admissible variation η. Inequality $\Phi''(0) \geq 0$ is equivalent to

$$\int_a^b \Big[L_{uu}[\hat{y}](t)(\eta^\sigma(t))^2 + 2L_{uv}[\hat{y}](t)\eta^\sigma(t) {}_a\Delta_h^\alpha \eta(t)$$

$$+ 2L_{uw}[\hat{y}](t)\eta^\sigma(t) {}_h\Delta_b^\beta \eta(t) + L_{vv}[\hat{y}](t)({}_a\Delta_h^\alpha \eta(t))^2 \qquad (12.82)$$

$$+ 2L_{vw}[\hat{y}](t) {}_a\Delta_h^\alpha \eta(t) {}_h\Delta_b^\beta \eta(t) + L_{ww}(t)({}_h\Delta_b^\beta \eta(t))^2 \Big] \Delta t \geq 0.$$

Let $\tau \in \mathbb{T}^{\kappa^2}$ be arbitrary, and choose $\eta : \mathbb{T} \to \mathbb{R}$ given by

$$\eta(t) = \begin{cases} h \text{ if } t = \sigma(\tau); \\ 0 \text{ otherwise.} \end{cases}$$

It follows that $\eta(a) = \eta(b) = 0$, i.e., η is an admissible variation. Using (12.62) we get

$$\int_a^b \Big[L_{uu}[\hat{y}](t)(\eta^\sigma(t))^2 + 2L_{uv}[\hat{y}](t)\eta^\sigma(t) {}_a\Delta_h^\alpha \eta(t) + L_{vv}[\hat{y}](t)({}_a\Delta_h^\alpha \eta(t))^2 \Big] \Delta t$$

$$= \int_a^b \Big[L_{uu}[\hat{y}](t)(\eta^\sigma(t))^2 + 2L_{uv}[\hat{y}](t)\eta^\sigma(t)$$

$$\times \left(h^\gamma \eta^\Delta(t) + \frac{\gamma}{\Gamma(\gamma+1)} \int_a^t (t + \gamma h - \sigma(s))_h^{(\gamma-1)} \eta^\Delta(s)\Delta s \right)$$

$$+ L_{vv}[\hat{y}](t) \left(h^\gamma \eta^\Delta(t) + \frac{\gamma}{\Gamma(\gamma+1)} \int_a^t (t + \gamma h - \sigma(s))_h^{(\gamma-1)} \eta^\Delta(s)\Delta s \right)^2 \Big] \Delta t$$

$$= h^3 L_{uu}[\hat{y}](\tau) + 2h^{\gamma+2} L_{uv}[\hat{y}](\tau) + h^{\gamma+1} L_{vv}[\hat{y}](\tau) + \int_{\sigma(\tau)}^b L_{vv}[\hat{y}](t)$$

$$\times \left(h^\gamma \eta^\Delta(t) + \frac{\gamma}{\Gamma(\gamma+1)} \int_a^t (t + \gamma h - \sigma(s))_h^{(\gamma-1)} \eta^\Delta(s)\Delta s \right)^2 \Delta t.$$

Observe that

$$\int_{\sigma(\tau)}^b L_{vv}[\hat{y}](t) \left(h^\gamma \eta^\Delta(t) + \frac{\gamma}{\Gamma(\gamma+1)} \int_a^t (t + \gamma h - \sigma(s))_h^{(\gamma-1)} \eta^\Delta(s)\Delta s \right)^2 \Delta t$$

$$= h^{2\gamma+1}(\gamma-1)^2 L_{vv}[\hat{y}](\sigma(\tau))$$

$$+ \int_{\sigma^2(\tau)}^b L_{vv}[\hat{y}](t) \left(\frac{\gamma}{\Gamma(\gamma+1)} \int_a^t (t + \gamma h - \sigma(s))_h^{(\gamma-1)} \eta^\Delta(s)\Delta s \right)^2 \Delta t.$$

Let $t \in [\sigma^2(\tau), \rho(b)] \cap h\mathbb{Z}$. Since

$$\frac{\gamma}{\Gamma(\gamma+1)} \int_a^t (t + \gamma h - \sigma(s))_h^{(\gamma-1)} \eta^\Delta(s) \Delta s$$

$$= \frac{\gamma}{\Gamma(\gamma+1)} \left[\int_a^{\sigma(\tau)} (t + \gamma h - \sigma(s))_h^{(\gamma-1)} \eta^\Delta(s) \Delta s \right.$$

$$\left. + \int_{\sigma(\tau)}^t (t + \gamma h - \sigma(s))_h^{(\gamma-1)} \eta^\Delta(s) \Delta s \right]$$

$$= h \frac{\gamma}{\Gamma(\gamma+1)} \left[(t + \gamma h - \sigma(\tau))_h^{(\gamma-1)} - (t + \gamma h - \sigma(\sigma(\tau)))_h^{(\gamma-1)} \right]$$

$$= \frac{\gamma h^\gamma}{\Gamma(\gamma+1)} \left[\frac{\left(\frac{t-\tau}{h} + \gamma - 1\right) \Gamma\left(\frac{t-\tau}{h} + \gamma - 1\right) - \left(\frac{t-\tau}{h}\right) \Gamma\left(\frac{t-\tau}{h} + \gamma - 1\right)}{\left(\frac{t-\tau}{h}\right) \Gamma\left(\frac{t-\tau}{h}\right)} \right]$$

$$= h^2 \frac{\gamma(\gamma-1)}{\Gamma(\gamma+1)} (t + \gamma h - \sigma(\sigma(\tau)))_h^{(\gamma-2)},$$

$$\tag{12.83}$$

we conclude that

$$\int_{\sigma^2(\tau)}^b L_{vv}[\hat{y}](t) \left(\frac{\gamma}{\Gamma(\gamma+1)} \int_a^t (t + \gamma h - \sigma(s))_h^{(\gamma-1)} \eta^\Delta(s) \Delta s \right)^2 \Delta t$$

$$= \int_{\sigma^2(\tau)}^b L_{vv}[\hat{y}](t) \left(h^2 \frac{\gamma(\gamma-1)}{\Gamma(\gamma+1)} (t + \gamma h - \sigma^2(\tau))_h^{(\gamma-2)} \right)^2 \Delta t.$$

Note that we can write $_t\Delta_b^\beta \eta(t) = -_h\Delta_{\rho(b)}^{-\nu} \eta^\Delta(t - \nu h)$ because $\eta(b) = 0$. It is not difficult to see that the following equality holds:

$$\int_a^b 2L_{uw}[\hat{y}](t) \eta^\sigma(t) {}_h\Delta_b^\beta \eta(t) \Delta t = -\int_a^b 2L_{uw}[\hat{y}](t) \eta^\sigma(t) {}_h\Delta_{\rho(b)}^{-\nu} \eta^\Delta(t - \nu h) \Delta t$$

$$= 2h^{2+\nu} L_{uw}[\hat{y}](\tau)(\nu - 1).$$

Moreover,

$$\int_a^b 2L_{vw}[\hat{y}](t) {}_a\Delta_h^\alpha \eta(t) {}_h\Delta_b^\beta \eta(t) \Delta t$$

$$= -2 \int_a^b L_{vw}[\hat{y}](t) \left\{ \left(h^\gamma \eta^\Delta(t) + \gamma \int_a^t \frac{(t + \gamma h - \sigma(s))_h^{(\gamma-1)} \eta^\Delta(s)}{\Gamma(\gamma+1)} \Delta s \right) \right.$$

$$\left. \times \left[h^\nu \eta^\Delta(t) + \frac{\nu}{\Gamma(\nu+1)} \int_{\sigma(t)}^b (s + \nu h - \sigma(t))_h^{(\nu-1)} \eta^\Delta(s) \Delta s \right] \right\} \Delta t$$

$$= 2h^{\gamma+\nu+1}(\nu - 1) L_{vw}[\hat{y}](\tau) + 2h^{\gamma+\nu+1}(\gamma - 1) L_{vw}[\hat{y}](\sigma(\tau)).$$

Finally, we have that

$$\int_a^b L_{ww}[\hat{y}](t)({}_h\Delta_b^\beta \eta(t))^2 \Delta t$$

$$= \int_a^{\sigma(\sigma(\tau))} L_{ww}[\hat{y}](t)$$

$$\times \left[h^\nu \eta^\Delta(t) + \frac{\nu}{\Gamma(\nu+1)} \int_{\sigma(t)}^b (s + \nu h - \sigma(t))_h^{(\nu-1)} \eta^\Delta(s)\Delta s \right]^2 \Delta t$$

$$= \int_a^\tau L_{ww}[\hat{y}](t) \left[\frac{\nu}{\Gamma(\nu+1)} \int_{\sigma(t)}^b (s + \nu h - \sigma(t))_h^{(\nu-1)} \eta^\Delta(s)\Delta s \right]^2 \Delta t$$

$$+ h L_{ww}[\hat{y}](\tau)(h^\nu - \nu h^\nu)^2 + h^{2\nu+1} L_{ww}[\hat{y}](\sigma(\tau))$$

$$= \int_a^\tau L_{ww}[\hat{y}](t)$$

$$\times \left[\frac{h\nu}{\Gamma(\nu+1)} \left\{ (\tau + \nu h - \sigma(t))_h^{(\nu-1)} - (\sigma(\tau) + \nu h - \sigma(t))_h^{(\nu-1)} \right\} \right]^2 \Delta t$$

$$+ h L_{ww}[\hat{y}](\tau)(h^\nu - \nu h^\nu)^2 + h^{2\nu+1} L_{ww}[\hat{y}](\sigma(\tau)).$$

Similarly as we did in (12.83), we can prove that

$$h\frac{\nu}{\Gamma(\nu+1)} \left\{ (\tau + \nu h - \sigma(t))_h^{(\nu-1)} - (\sigma(\tau) + \nu h - \sigma(t))_h^{(\nu-1)} \right\}$$

$$= h^2 \frac{\nu(1-\nu)}{\Gamma(\nu+1)}(\tau + \nu h - \sigma(t))_h^{(\nu-2)}.$$

Thus, we have that inequality (12.82) is equivalent to

$$h\left\{ h^2 L_{uu}[\hat{y}](t) + 2h^{\gamma+1} L_{uv}[\hat{y}](t) + h^\gamma L_{vv}[\hat{y}](t) + L_{vv}(\sigma(t))(\gamma h^\gamma - h^\gamma)^2 \right.$$

$$+ \int_{\sigma(\sigma(t))}^b h^3 L_{vv}(s) \left(\frac{\gamma(\gamma-1)}{\Gamma(\gamma+1)}(s + \gamma h - \sigma(\sigma(t)))_h^{(\gamma-2)} \right)^2 \Delta s$$

$$+ 2h^{\nu+1} L_{uw}[\hat{y}](t)(\nu-1) + 2h^{\gamma+\nu}(\nu-1) L_{vw}[\hat{y}](t)$$

$$+ 2h^{\gamma+\nu}(\gamma-1) L_{vw}(\sigma(t)) + h^{2\nu} L_{ww}[\hat{y}](t)(1-\nu)^2 + h^{2\nu} L_{ww}[\hat{y}](\sigma(t))$$

$$\left. + \int_a^t h^3 L_{ww}[\hat{y}](s) \left(\frac{\nu(1-\nu)}{\Gamma(\nu+1)}(t + \nu h - \sigma(s))^{\nu-2} \right)^2 \Delta s \right\} \geq 0. \quad (12.84)$$

Because $h > 0$, (12.84) is equivalent to (12.81). The theorem is proved. □

The next result is a simple corollary of Theorem 12.16 and can be found in (Bohner, 2004, Result 1.3).

Corollary 12.4 (The *h*-Legendre necessary condition). *Let* $\mathbb{T}$ *be the time scale* $h\mathbb{Z}$, $h > 0$, *with the forward jump operator* σ *and the delta derivative* Δ. *Assume* $a, b \in \mathbb{T}$, $a < b$. *If* $\hat{y}$ *is a solution to the problem*

$$\mathcal{L}[y] = \int_a^b L(t, y^\sigma(t), y^\Delta(t))\Delta t \longrightarrow \min,$$

$$y(a) = A, \quad y(b) = B,$$

then the inequality

$$h^2 L_{uu}[\hat{y}](t) + 2hL_{uv}[\hat{y}](t) + L_{vv}[\hat{y}](t) + L_{vv}[\hat{y}](\sigma(t)) \geq 0 \qquad (12.85)$$

holds for all $t \in \mathbb{T}^{\kappa^2}$, *where* $[\hat{y}](t) = (t, \hat{y}^\sigma(t), \hat{y}^\Delta(t))$.

Proof. Choose $\alpha = 1$ and a Lagrangian L that does not depend on w. Then, $\gamma = 0$ and the result follows immediately from Theorem 12.16. $\square$

Remark 12.21. When h goes to zero we have $\sigma(t) = t$ and inequality (12.85) coincides with Legendre's classical necessary optimality condition $L_{vv}[\hat{y}](t) \geq 0$ (cf., e.g., (van Brunt, 2004)).

12.6.5 *Examples*

In this section we present some illustrative examples.

Example 12.12. Let us consider the following problem:

$$\mathcal{L}[y] = \frac{1}{2} \int_0^1 \left({}_0\Delta_h^{\frac{3}{4}} y(t)\right)^2 \Delta t \longrightarrow \min, \qquad (12.86)$$

$$y(0) = 0, \quad y(1) = 1.$$

We consider (12.86) with different values of h. Numerical results show that when h tends to zero the h-fractional Euler–Lagrange extremal tends to the fractional continuous extremal: when $h \to 0$ (12.86) tends to the fractional continuous variational problem in the Riemann–Liouville sense studied in (Agrawal, 2002, Example 1), with solution given by

$$y(t) = \frac{1}{2} \int_0^t \frac{dx}{[(1-x)(t-x)]^{\frac{1}{4}}}. \qquad (12.87)$$

This is illustrated in Fig. 12.5.

In this example for each value of h there is a unique h-fractional Euler–Lagrange extremal, solution of (12.71), which always verifies the h-fractional Legendre necessary condition (12.81).

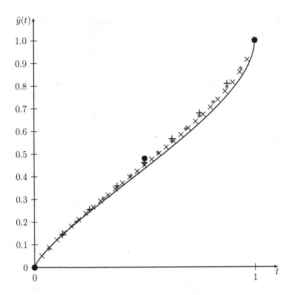

Fig. 12.5   Extremal $\tilde{y}(t)$ for problem of Example 12.12 with different values of h: $h = 0.50$ (•); $h = 0.125$ (+); $h = 0.0625$ (∗); $h = 1/30$ (×). The continuous line represents function (12.87).

Example 12.13. Let us consider the following problem:

$$\mathcal{L}[y] = \int_0^1 \left[\frac{1}{2} \left({}_0\Delta_h^\alpha y(t) \right)^2 - y^\sigma(t) \right] \Delta t \longrightarrow \min, \tag{12.88}$$

$$y(0) = 0, \quad y(1) = 0.$$

We begin by considering problem (12.88) with a fixed value for α and different values of h. The extremals $\tilde{y}$ are obtained using the Euler–Lagrange equation (12.71). As in Example 12.12 the numerical results show that when h tends to zero the extremal of the problem tends to the extremal of the corresponding continuous fractional problem of the calculus of variations in the Riemann–Liouville sense. More precisely, when h approximates zero, problem (12.88) tends to the fractional continuous problem studied in (Agrawal, 2008b, Example 2). For $\alpha = 1$ and $h \to 0$ the extremal of (12.88) is given by $y(t) = \frac{1}{2}t(1-t)$, which coincides with the extremal of the classical problem of the calculus of variations

$$\mathcal{L}[y] = \int_0^1 \left(\frac{1}{2} y'(t)^2 - y(t) \right) dt \longrightarrow \min,$$

$$y(0) = 0, \quad y(1) = 0.$$

This is illustrated in Fig. 12.6 for $h = \frac{1}{2^i}$, $i = 1, 2, 3, 4$.

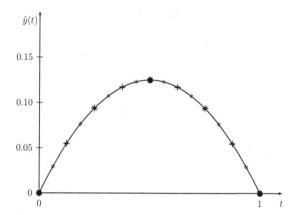

Fig. 12.6 Extremal $\tilde{y}(t)$ for problem (12.88) with $\alpha = 1$ and different values of h: $h = 0.5$ ($\bullet$); $h = 0.25$ ($\times$); $h = 0.125$ ($+$); $h = 0.0625$ ($*$).

In this example, for each value of α and h, we only have one extremal (we only have one solution to (12.71) for each α and h). The Legendre condition (12.81) is always verified along the extremals. Figure 12.7 shows the extremals of problem (12.88) for a fixed value of h ($h = 1/20$) and different values of α. The numerical results show that when α tends to one, the extremal tends to the solution of the classical (integer-order) discrete-time problem.

Our last example shows that the h-fractional Legendre necessary optimality condition can be a very useful tool. In Example 12.14 we consider a problem for which the h-fractional Euler–Lagrange equation gives several candidates but just a few of them verify the Legendre condition (12.81).

Example 12.14. Let us consider the following problem:

$$\mathcal{L}[y] = \int_a^b \left({_a\Delta_h^\alpha y(t)} \right)^3 + \theta \left({_h\Delta_b^\alpha y(t)} \right)^2 \Delta t \longrightarrow \min,$$

$$y(a) = 0, \quad y(b) = 1.$$

(12.89)

For $\alpha = 0.8$, $\beta = 0.5$, $h = 0.25$, $a = 0$, $b = 1$, and $\theta = 1$, problem (12.89) has eight different Euler–Lagrange extremals. As we can see on Table 12.4

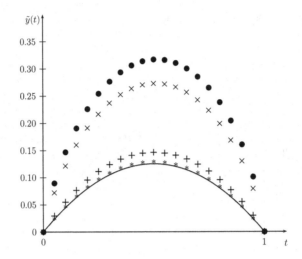

Fig. 12.7   Extremal $\tilde{y}(t)$ for (12.88) with $h = 0.05$ and different values of α: $\alpha = 0.70$ ($\bullet$); $\alpha = 0.75$ ($\times$); $\alpha = 0.95$ (+); $\alpha = 0.99$ ($*$). The continuous line is $y(t) = \frac{1}{2}t(1-t)$.

only two of the candidates verify the Legendre condition. To determine the best candidate we compare the values of the functional $\mathcal{L}$ along the two good candidates. The extremal we are looking for is given by the candidate number five on Table 12.4.

Table 12.4   There exist eight Euler–Lagrange extremals for problem (12.89) with $\alpha = 0.8$, $\beta = 0.5$, $h = 0.25$, $a = 0$, $b = 1$, and $\theta = 1$, but only two of them satisfy the fractional Legendre condition (12.81).

#	$\tilde{y}(1/4)$	$\tilde{y}(1/2)$	$\tilde{y}(3/4)$	$\mathcal{L}[\tilde{y}]$	condition (12.81)
1	-0.5511786	0.0515282	0.5133134	9.3035911	Not verified
2	0.2669091	0.4878808	0.7151924	2.0084203	Verified
3	-2.6745703	0.5599360	-2.6730125	698.4443232	Not verified
4	0.5789976	1.0701515	0.1840377	12.5174960	Not verified
5	1.0306820	1.8920322	2.7429222	-32.7189756	Verified
6	0.5087946	-0.1861431	0.4489196	10.6730959	Not verified
7	4.0583690	-1.0299054	-5.0030989	2451.7637948	Not verified
8	-1.7436106	-3.1898449	-0.8850511	238.6120299	Not verified

For problem (12.89) with $\alpha = 0.3$, $h = 0.1$, $a = 0$, $b = 0.5$, and $\theta = 0$, we obtain the results of Table 12.5: there exist sixteen Euler–Lagrange extremals but only one satisfy the fractional Legendre condition. The extremal we are looking for is given by the candidate number six on Table 12.5.

Table 12.5 There exist sixteen Euler–Lagrange extremals for problem (12.89) with $\alpha = 0.3$, $h = 0.1$, $a = 0$, $b = 0.5$, and $\theta = 0$, but only one (candidate #6) satisfy the fractional Legendre condition (12.81).

#	$\tilde{y}(0.1)$	$\tilde{y}(0.2)$	$\tilde{y}(0.3)$	$\tilde{y}(0.4)$	$\mathcal{L}[\tilde{y}]$	(12.81)
1	-0.305570704	-0.428093486	0.223708338	0.480549114	12.25396	No
2	-0.427934654	-0.599520948	0.313290997	-0.661831134	156.2318	No
3	0.284152257	-0.227595659	0.318847274	0.531827387	8.669646	No
4	-0.277642565	0.222381632	0.386666793	0.555841555	6.993518	No
5	0.387074742	-0.310032839	0.434336603	-0.482903047	110.7913	No
6	0.259846344	0.364035314	0.463222456	0.597907505	5.104389	Yes
7	-0.375094681	0.300437245	0.522386246	-0.419053781	93.95317	No
8	0.343327771	0.480989769	0.61204299	-0.280908953	69.23498	No
9	0.297792192	0.417196073	-0.218013689	0.460556635	14.12228	No
10	0.41283304	0.578364133	-0.302235104	-0.649232892	157.8273	No
11	-0.321401682	0.257431098	-0.360644857	0.400971272	19.87469	No
12	0.330157414	-0.264444122	-0.459803086	0.368850105	24.84476	No
13	-0.459640837	0.368155651	-0.515763025	-0.860276767	224.9965	No
14	-0.359429958	-0.50354835	-0.640748011	0.294083676	34.43516	No
15	0.477760586	-0.382668914	-0.66536683	-0.956478654	263.3075	No
16	-0.541587541	-0.758744525	-0.965476394	-1.246195157	392.9592	No

12.6.6 *Conclusion*

In this section we introduce a new fractional difference variational calculus in the time scale $(h\mathbb{Z})_a$, $h > 0$ and a a real number, for Lagrangians depending on left and right discrete-time fractional derivatives. Our objective was to introduce the concept of left and right fractional sum/difference (cf. Definition 12.21) and to develop the theory of fractional difference calculus. An Euler–Lagrange type equation (12.71), fractional natural boundary conditions (12.79) and (12.80), and a second-order Legendre type necessary optimality condition (12.81), were obtained. The results are based on a new discrete fractional summation by parts formula (12.68) for $(h\mathbb{Z})_a$. Obtained first and second-order necessary optimality conditions were implemented computationally in the computer algebra systems Maple® and Maxima (the Maxima code can be found in (Bastos, 2012)). The numerical results show that solutions to the considered fractional problems become the classical discrete-time solutions when the fractional-order of the discrete-derivatives are integer values, and that they converge to the fractional continuous-time solutions when h tends to zero. The Legendre type condition is useful to eliminate false candidates identified via the Euler–Lagrange fractional equation. The results of the chapter are formulated using standard notations of the theory of time scales (Bohner and Peterson, 2001; Jarad, Baleanu and Maraaba, 2008; Malinowska and Torres, 2009). Undoubtedly, much remains to be done in the development of the theory of discrete fractional calculus of variations in $(h\mathbb{Z})_a$ initiated in (Bastos, Ferreira and Torres, 2011b) and here presented. In particular, a generalization of this chapter's results to an arbitrary time scale $\mathbb{T}$ is still an open question. The reader interested in the development of the theory presented in this chapter, with handy tools for the explicit solution of discrete equations involving left and right fractional difference operators, is referred to (Ferreira and Torres, 2011).

Chapter 13

Conclusion

The realm of numerical methods in scientific fields is vastly growing due to the very fast progresses in computational sciences and technologies. Nevertheless, the intrinsic complexity of fractional calculus, caused partially by non-local properties of fractional derivatives and integrals, makes it rather difficult to find efficient numerical methods in this field. It seems enough to mention here that, up to the time of this book, and to the best of our knowledge, there is no routine available for solving a fractional differential equation as there is the Runge–Kutta method for ordinary ones. Despite this fact, however, the literature exhibits a growing interest and improving achievements in numerical methods for fractional calculus in general and fractional variational problems specifically.

This book is devoted to the discussion of some aspects of the very well-known methods for solving variational problems. Namely, we studied the notions of direct and indirect methods in the classical calculus of variations and also we mentioned some connections to optimal control. Consequently, we introduced the generalizations of these notions to the field of fractional calculus of variations and fractional optimal control.

The method of finite differences, as discussed here, seems to be a potential first candidate to solve fractional variational problems. Although a first order approximation was used for all examples, the results are satisfactory and even though it is more complicated than in the classical case, it still inherits some sort of simplicity and an ease of implementation. Roughly speaking, an Euler-like direct method reduces a variational problem to the solution of a system of algebraic equations. When the system is linear, we can freely increase the number of mesh points, n, and obtain better solutions as long as the resulted matrix of coefficients is invertible. The method is very fast, in this case, and the execution time is of order 10^{-4} seconds for Examples 8.1 and 8.2. It is worth, however, to keep in mind that the Grünwald–Letnikov approximation is of first order, $\mathcal{O}(h)$, and even a large n cannot result in a high precision. Actually, by increasing n, the solution slowly converges and in Example 8.2, a grid of 30 points has the same order of error, 10^{-3}, as a 5 points grid. The situation is completely different when the problem ends with a nonlinear system. In Example 8.3, a small number of mesh points, $n = 5$, results in a poor solution with the error $E = 1.4787$. The MATLAB® built in function fsolve takes 0.0126 seconds to solve the problem. As one increases the number of mesh points, the solution gets closer to the analytic solution and the required time increases drastically. Finally, by $n = 90$ we have $E = 0.0618$ and the time is $T = 26.355$ seconds.

In practice, we have no idea about the solution in advance and the worst case should be taken into account. Comparing the results of the three examples considered, reveals that for a typical fractional variational problem, the Euler-like direct method needs a large number of mesh points and most likely a long running time.

The lack of efficient numerical methods for fractional variational problems, is overcome partially by the indirect methods of this book. Once we transformed the fractional variational problem to an approximated classical one, the majority of classical methods can be applied to get an approximate solution. Nevertheless, the procedure is not completely straightforward. The singularity of fractional operators is still present in the approximating formulas and it makes the solution procedure more complicated.

During the last three decades, several numerical methods have been developed in the field of fractional calculus. Some of their advantages, disadvantages, and improvements, are given in (Aoun *et al.*, 2004). Based on two continuous expansion formulas (6.2) and (6.7) for the left Riemann–Liouville fractional derivative, we studied two approximations (6.4) and (6.10) and their applications in the computation of fractional derivatives. Despite the fact that the approximation (6.4) encounters some difficulties from the presence of higher-order derivatives, it exhibits better results at least for the evaluation of fractional derivatives. The same studies were carried out for fractional integrals as well as some other fractional operators, namely Hadamard derivatives and integrals, and Caputo derivatives. The full details regarding these approximations and their advantages, disadvantages and applications can be found in (Pooseh, Almeida and Torres, 2013a, 2012a,c).

Approximation (6.10) can also be generalized to include higher-order derivatives in the form of (6.15). The possibility of using (6.10) to compute fractional derivatives for a set of tabular data was discussed. Fractional differential equations are also treated successfully. In this case the lack of initial conditions makes (6.4) less useful. In contrast, one can freely increase N, the order of approximation (6.10), and find better approximations. Comparing with (6.13), our modification provides better results.

For fractional variational problems, the proposed expansions may be used at two different stages during the solution procedure. The first approach, the one considered in Chapter 8.3.2, consists in a direct approximation of the problem, and then treating it as a classical problem, using standard methods to solve it. The second approach, Section 10.4.1, is to apply the fractional Euler–Lagrange equation and then to use the approximations in order to obtain a classical differential equation. The results concerning the application of the approximations proposed in this work have been published in (Almeida, Pooseh and Torres, 2012; Pooseh, Almeida and Torres, 2013c,d, 2014).

The direct methods for fractional variational problems presented in this book, can be improved in some stages. One can try different approximations for the fractional derivative that exhibit higher-order precisions, e.g. Diethelm's backward finite differences (Diethelm *et al.*, 2005). Better quadrature rules can be applied to discretize the functional and, finally, we can apply more sophisticated algorithms for solving the resulting system of algebraic equations. Further works are needed to cover different types of fractional variational problems.

Regarding indirect methods, the idea of transforming a fractional problem to a classic one seems a useful way of extending the available classic methods to the field of fractional variational problems. Nevertheless, improvements are needed to avoid the singularities of the approximations (6.10) and (6.4). A more practical goal is to implement some software packages or tools to solve certain classes of fractional variational problems. Following this research direction may also end in some solvers for fractional differential equations.

The theory developed in this book can be very useful in applications. One such example of application can be found in epidemiology (Pooseh, Rodrigues and Torres, 2011), where the approach developed in this book is illustrated with an outbreak of dengue disease, motivated by the first dengue epidemic ever recorded in the Cape Verde islands off the coast of West Africa, in 2009. Describing the reality through a mathematical model, usually a system of differential equations, is a hard task that has an inherent compromise between simplicity and accuracy. In (Pooseh, Rodrigues and Torres, 2011), a very basic model to dengue epidemics is considered. It turns out that, in general, this basic/classical model does not provide enough good results. In order to have better results, that fit the reality, a more specific and complicated set of differential equations have been investigated in the literature, see (Rodrigues, Monteiro and Torres, 2010a,b; Rodrigues *et al.*, 2012) and references therein. A new approach to the subject is introduced in (Pooseh, Rodrigues and Torres, 2011), where the simple model is kept and the usual (local) derivatives are substituted by (non-local) fractional differentiation. The use of fractional derivatives allow one to model memory effects, and result in a more powerful approach to epidemiological models: one can then design the order α of fractional differentiation that best corresponds to reality. The classical case is recovered by taking the limit when α goes to one. Such investigations show that even a simple fractional model may give surprisingly good results (Pooseh, Rodrigues and Torres, 2011). However, the transformation of a classical model into a fractional one makes it very sensitive to the order of differentiation α: a small change in α may result in a big change in the final result, and a more sophisticated study has been reported in (Diethelm, 2013).

We ended the book by considering fractional variational problems of variable order and variational problems defined by fractional difference operators in the time scales $\mathbb{T} = \mathbb{Z}$ and $\mathbb{T} = (h\mathbb{Z})_a$, where much remains to be done. We hope the present book will potentiate further research in the beautiful and useful subject of the fractional calculus of variations.

Appendix A

MATLAB® code

In this appendix we provide some MATLAB® code used in the examples. The scripts are a straightforward implementation of the theory presented throughout the book. Only basic MATLAB® features are used, and we trust the reader will have no difficulty to adapt the code to her/his preferred programming language or computer software.

Example 8.1

```
t=linspace(0,1,100);
fig=figure;
plot(t,t.^2,'r')

linstyle={'--','-.m','.'};
str=cell(1,3);
for iter=1:3
    n=5*factorial(iter);
    t=linspace(0,1,n);
    h=t(2)-t(1);
    A=0*t;
    for i=1:n
        A(i)=gamma(i-1-.5)*h^(1.5)/gamma(-.5)/gamma(i);
    end
    M=zeros(n-2,n-2);
    for i=1:n-2
        p=0;
        for j=i:n-2
            M(i,j)=0;
            for k=1:n-j
                M(i,j)=M(i,j)+A(k+p)*A(k);
            end
```

```
             p=p+1;
        end
    end

    for j=1:n-3
        p=1;
        for i=j+1:n-2
            M(i,j)=0;
            for k=1:n-i
                M(i,j)=M(i,j)+A(k)*A(k+p);
            end
            p=p+1;
        end
    end
    B=2*h*h/gamma(2.5);
    x1=0;
    xn=1;
    b=zeros(n-2,1);
    p=1;
    for i=1:n-2
        b(i)=-A(n-i)*A(1)*xn;
        for j=1:n-i
            b(i)=b(i)-A(j)*A(j+p)*x1+A(j)*B*t(j+p)^(1.5);
        end
        p=p+1;
    end
    x=M\b;
    hold on
    plot(t,[0;x;1],linstyle{iter})
    y=[0 x' 1];
    norm=max(abs(y-t.^2));
    str{iter}=['Approximation: n = ' num2str(n,'% -2d') ', ...
        Error= ' num2str(norm,1)];
end
legend('Analytic solution',str{1},str{2},str{3},2)
xlabel('t')
ylabel('x(t)')
```

Example 8.2

```
t=linspace(0,1,100);
g=1/2/gamma(2.5);
y=-g*(1-t).^1.5+(1-g)*t+g;
fig=figure;
plot(t,y,'k')
hold on
linstyle={'--k','-.k','.k'};
str={};
```

```
for iter=1:3
    n=5*factorial(iter);
    t=linspace(0,1,n);
    h=t(2)-t(1);
    A=0*t;
    for i=1:n
        A(i)=gamma(i-1-.5)*h^(.5)/gamma(-.5)/gamma(i);
    end
    M=2*eye(n-2,n-2);
    for i=1:n-3
        M(i,i+1)=-1;
        M(i+1,i)=-1;
    end
    x1=0;
    xn=1;
    b=zeros(n-2,1);
    for i=1:n-2
        b(i)=0;
        for j=1:n-i
            b(i)=b(i)+A(j);
        end
        b(i)=h*b(i)/2;
    end
    b(n-2)=b(n-2)+1;
    x=M\b;
    disp(toc)
    plot(t,[0;x;1],linstyle{iter})
    y=[0 x' 1];
    yt=-g*(1-t).^1.5+(1-g)*t+g;
    norm=max(abs(y-yt))
    str{iter}=['Approximation: n = ' num2str(n,'% -2d') ', ...
        Error= ' num2str(norm,1)];
end
legend('Analytic solution',str{1},str{2},str{3},2)
xlabel('t')
ylabel('x(t)')
```

Example 8.3

```
global t a;
a=.5;
n=5;
t=linspace(0,1,n);
fplot('16*t.^5-20*t.^3+5*t',[0 1],'r');
hold on
options=optimset('TolFun',1e-9,'TolX',1e-9);
linStyle={'--+','-.m','.'};
str=cell(1,3);
```

```
N=[5 20 60];
for i=1:3
    n=N(i);
    t=linspace(0,1,n);
    x0=0*t(2:n-1);
    x = fsolve(@Fun,x0,options);
    Time=toc
    el=max(abs([0 x 1]-16*t.^5+20*t.^3-5*t));
    str{i}=['Approximation: n = ' num2str(n,'% -2d') ', Error= ...
        ' num2str(el,3)];
    plot(t,[0 x 1], linStyle{i})
end
legend('Analytic solution',str{1},str{2},str{3},3)
xlabel('t')
ylabel('x(t)')

function  F=Fun(x)
global t a;
h=t(2)-t(1);
n=length(t);
F=0*x;
X=[0 x 1];
w=zeros(1,n+2);
for k=1:n
    w(k)=h^(-a)*gamma(k-1-a)/gamma(-a)/gamma(k);
end
c1=16*gamma(6)/gamma(5.5);
c2=20*gamma(4)/gamma(3.5);
c3=5*gamma(2)/gamma(1.5);
for j=1:length(x)
    F(j)=0;
    for i=j+1:n
        A=0;
        for k=1:i
            A=A+w(k)*X(i-k+1);
        end
        F(j)=F(j)+4*w(i-j)*(A-c1*t(i)^4.5+c2*t(i)^2.5-c3*t(i)^.5)^3;
    end
end
```

Example 10.1

```
function exm101
global a N EP;
EP=0.0001;
t0=0;
a=.5;
N=2;
```

```
options = [];%bvpset('FJacobian',@Jac,'BCJacobian',@BCJac);
solinit = bvpinit(linspace(t0,1,30),.1*[1 1 1 1 ]);
sol = bvp4c(@dEqs,@bcEq,solinit,options);
[A B C]=coefs(a,N);
t=sol.x;
u=(a+2)./t.*sol.y(1,:)-sol.y(3,:)./(2*t.^2.*(1+B*t.^(1-a)));
u(1)=0;
J=trapz(sol.x,(t.*u-(a+2)*sol.y(1,:)).^2);
solExact=(2/gamma(a+3))*sol.x.^(a+2);
plot(sol.x,solExact,'k')
hold on
plot(sol.x,sol.y(1,:),'k-.');
norm=max(abs(solExact-sol.y(1,:)));
legend('Exact solution: J=0',['Approximation: N=2, J=' ...
    num2str(J) ', Error=' num2str(norm)],2)
xlabel('t')
ylabel('x(t)')
figure
uExact=(2/gamma(a+2))*sol.x.^(a+1);
plot(sol.x,uExact,'k')
hold on
plot(sol.x,u,'k-.');
normu=max(abs(uExact-u));
legend('Exact solution',['Approximation: N=2 ', Error=' ...
    num2str(normu)],2)
xlabel('t')
ylabel('u(t)')

function dy=dEqs(t,y)
global a N EP;
[A B C]=coefs(a,N);
dy=zeros(2*N,1);
if t≤EP
    t=EP;
end
dy(1)=(-y(3)/2/t^2/(1+B*t^(1-a))+((a+2)/t-A*t^(-a))*y(1)
        +C(2)*t^(-1-a)*y(2)+t^2)/(1+B*t^(1-a)));
dy(2)=-y(1);
dy(3)=(-(a+2)/t+A*t^(-a))*y(3)/(1+B*t^(1-a))+y(4);
dy(4)=-C(2)*t^(-1-a)*y(3)/(1+B*t^(1-a));

function dFdy = Jac(t,y)
global a N EP;
[A B C]=coefs(a,N);
if t≤EP
    t=EP;
end
dFdy=[((a+2)/t-A*t^(-a))/(1+B*t^(1-a)), ...
    C(2)*t^(-1-a)/(1+B*t^(1-a)), -1/2/t^2/(1+B*t^(1-a))^2, 0
            -1, 0, 0, 0
```

```
       0, 0, (-(a+2)/t+A*t^(-a))/t/(1+B*t^(1-a)), -1
       0, 0, -C(2)*t^(-1-a)/(1+B*t^(1-a)), 0];

function res=bcEq(ya,yb)
global a;
res = [ya(1);
       ya(2);
       yb(1)-2/gamma(a+3);
       yb(4)];

function [dBCdya,dBCdyb] = BCJac(ya,yb)
dBCdya = [ 1, 0, 0, 0
           0, 1, 0, 0
           0, 0, 0, 0
           0, 0, 0, 0];

dBCdyb = [ 0, 0, 0, 0
           0, 0, 0, 0
           1, 0, 0, 0
           0, 0, 0, 1];

function [A,B,C]=coefs(a,N)
A1=1/gamma(1-a);
A2=1/gamma(2-a)/gamma(a-1);
A3=0;
for p=2:N
    A3=A3+gamma(p-1+a)/factorial(p-1);
end
A=A1-A2*A3;
B1=1/gamma(2-a);
B2=0;
for p=1:N
    B2=B2+gamma(p-1+a)/gamma(a-1)/factorial(p);
end
B=B1+B1*B2;
C=zeros(1,N);
for p=2:N
    C(p)=A2*gamma(p-1+a)/gamma(p);
end
```

Example 10.2

```
global a N EP;
EP=0.0001;
t0=0;
a=.5;
N=2;
options = bvpset('FJacobian',@Jac,'BCJacobian',@BCJac);
```

```
solinit = bvpinit(linspace(t0,1,30),.1*[1 1 1 1 ],.5);
sol = bvp5c(@dEqs,@bcEq,solinit,options);
 [A B C]=coefs(a,N);
 u=.5*B^(-1)*sol.x.^(a-1).*sol.y(3,:)+1;
 u(1)=0;
 J=trapz(sol.x,(u-1).^2);
sol.parameters
t=sol.parameters*sol.x;
plot(t,sol.y(1,:),'k');
hold on
plot(t,sol.y(3,:));
legend('Analytic: J=0',['Approximation: N=2, J=' num2str(J)],2)
xlabel('t')
ylabel('x(t)')


function dy=dEqs(t,y,T)
global a N EP;
[A B C]=coefs(a,N);
dy=zeros(2*N,1);
if t≤EP
    t=EP;
end
dy=T*[(-y(3)/(1+B*(t*T)^(1-a))+(1-A*(t*T)^(-a))*y(1)
        +C(2)*(t*T)^(-1-a)*y(2)
        +(t*T)^(-a)/gamma(1-a))/(1+B*(t*T)^(1-a));
    -y(1);
    (-1+A*(t*T)^(-a))*y(3)/(1+B*(t*T)^(1-a))+y(4);
    -C(2)*(t*T)^(-1-a)*y(3)/(1+B*(t*T)^(1-a))];

function dFdy = Jac(t,y)
global a N EP;
[A B C]=coefs(a,N);
if t≤EP
    t=EP;
end
dFdy = ...
    [((a+2)/t-A*t^(-a))/(1+B*t^(1-a)),C(2)*t^(-1-a)/(1+B*t^(1-a)),
                -1/2/t^2/(1+B*t^(1-a))^2,  0
            -1, 0, 0, 0
             0, 0, (-(a+2)/t+A*t^(-a))/t/(1+B*t^(1-a)), -1
             0, 0, -C(2)*t^(-1-a)/(1+B*t^(1-a)), 0];

function res=bcEq(ya,yb,T)
global a N;
[A B C]=coefs(a,N);
Trans=(-yb(3)/(1+B*T^(1-a))+(1-A*T^(-a))*yb(1)+C(2)*T^(-1-a)*yb(2)+...
            T^(-a)/gamma(1-a))/(1+B*T^(1-a));
res = [ya(1)-1;
        ya(2);
```

```
      yb(1)-2;
      yb(4);
      Trans*yb(3)-yb(1)*yb(4)];

function [dBCdya,dBCdyb] = BCJac(ya,yb)
dBCdya = [ 1, 0, 0, 0
           0, 1, 0, 0
           0, 0, 0, 0
           0, 0, 0, 0];

dBCdyb = [ 0, 0, 0, 0
           0, 0, 0, 0
           1, 0, 0, 0
           0, 0, 0, 1];

function [A,B,C]=coefs(a,N)
A1=1/gamma(1-a);
A2=1/gamma(2-a)/gamma(a-1);
A3=0;
for p=2:N
    A3=A3+gamma(p-1+a)/factorial(p-1);
end
A=A1-A2*A3;
B1=1/gamma(2-a);
B2=0;
for p=1:N
    B2=B2+gamma(p-1+a)/gamma(a-1)/factorial(p);
end
B=B1+B1*B2;
C=zeros(1,N);
for p=2:N
    C(p)=A2*gamma(p-1+a)/gamma(p);
end
```

Bibliography

R. Agarwal, M. Bohner, D. O'Regan and A. Peterson, Dynamic equations on time scales: a survey, J. Comput. Appl. Math. **141** (2002), no. 1-2, 1–26.

O.P. Agrawal, Formulation of Euler-Lagrange equations for fractional variational problems, J. Math. Anal. Appl. **272** (2002), no. 1, 368–379.

O.P. Agrawal, A general formulation and solution scheme for fractional optimal control problems, Nonlinear Dynam. **38** (2004), no. 1-4, 323–337.

O.P. Agrawal, Fractional variational calculus and the transversality conditions, J. Phys. A **39** (2006), no. 33, 10375–10384.

O.P. Agrawal, Fractional variational calculus in terms of Riesz fractional derivatives, J. Phys. A **40** (2007a), no. 24, 6287–6303.

O.P. Agrawal, Generalized Euler-Lagrange equations and transversality conditions for FVPs in terms of the Caputo derivative, J. Vib. Control **13** (2007b), no. 9-10, 1217–1237.

O.P. Agrawal, A formulation and numerical scheme for fractional optimal control problems, J. Vib. Control **14** (2008a), no. 9-10, 1291–1299.

O.P. Agrawal, A general finite element formulation for fractional variational problems, J. Math. Anal. Appl. **337** (2008b), no. 1, 1–12.

O.P. Agrawal, O. Defterli and D. Baleanu, Fractional optimal control problems with several state and control variables, J. Vib. Control **16** (2010), no. 13, 1967–1976.

O.P. Agrawal, S.I. Muslih and D. Baleanu, Generalized variational calculus in terms of multi-parameters fractional derivatives, Commun. Nonlinear Sci. Numer. Simul. **16** (2011), no. 12, 4756–4767.

E. Akin, Cauchy functions for dynamic equations on a measure chain, J. Math. Anal. Appl. **267** (2002), no. 1, 97–115.

R. Almeida, Fractional variational problems with the Riesz-Caputo derivative, Appl. Math. Lett. **25** (2012), no. 2, 142–148.

R. Almeida, General necessary conditions for infinite horizon fractional variational problems, Appl. Math. Lett. **26** (2013), no. 7, 787–793.

R. Almeida, R.A.C. Ferreira and D.F.M. Torres, Isoperimetric problems of the calculus of variations with fractional derivatives, Acta Math. Sci. Ser. B Engl. Ed. **32** (2012), no. 2, 619–630.

R. Almeida, H. Khosravian-Arab and M. Shamsi, A generalized fractional variational problem depending on indefinite integrals: Euler-Lagrange equation and numerical solution, J. Vib. Control **19** (2013), no. 14, 2177–2186.

R. Almeida and A.B. Malinowska, Generalized transversality conditions in fractional calculus of variations, Commun. Nonlinear Sci. Numer. Simul. **18** (2013), no. 3, 443–452.

R. Almeida, A.B. Malinowska and D.F.M. Torres, A fractional calculus of variations for multiple integrals with application to vibrating string, J. Math. Phys. **51** (2010), no. 3, 033503, 12 pp.

R. Almeida, A.B. Malinowska and D.F.M. Torres, Fractional Euler-Lagrange differential equations via Caputo derivatives, in *Fractional Dynamics and Control* (eds: D. Baleanu, J. A. Tenreiro Machado, and A. Luo), Springer New York, 2012, Part 2, 109–118.

R. Almeida, S. Pooseh and D.F.M. Torres, Fractional variational problems depending on indefinite integrals, Nonlinear Anal. **75** (2012), no. 3, 1009–1025.

A. Almeida and S. Samko, Fractional and hypersingular operators in variable exponent spaces on metric measure spaces, Mediterr. J. Math. **6** (2009), no. 2, 215–232.

R. Almeida and D.F.M. Torres, Calculus of variations with fractional derivatives and fractional integrals, Appl. Math. Lett. **22** (2009a), no. 12, 1816–1820.

R. Almeida and D.F.M. Torres, Hölderian variational problems subject to integral constraints, J. Math. Anal. Appl. **359** (2009b), no. 2, 674–681.

R. Almeida and D.F.M. Torres, Isoperimetric problems on time scales with nabla derivatives, J. Vib. Control **15** (2009c), no. 6, 951–958.

R. Almeida and D.F.M. Torres, Leitmann's direct method for fractional optimization problems, Appl. Math. Comput. **217** (2010), no 3, 956–962.

R. Almeida and D.F.M. Torres, Necessary and sufficient conditions for the fractional calculus of variations with Caputo derivatives, Commun. Nonlinear Sci. Numer. Simulat. **16** (2011), no. 3, 1490–1500.

R. Almeida and D.F.M. Torres, An expansion formula with higher-order derivatives for fractional operators of variable order, The Scientific World Journal **2013** (2013), Art. ID 915437, 11 pp.

G.A. Anastassiou, *Intelligent mathematics: computational analysis*, Intelligent Systems Reference Library, 5, Springer, Berlin, 2011.

G.E. Andrews, R. Askey and R. Roy, *Special functions*, Encyclopedia of Mathematics and its Applications, 71, Cambridge University Press, Cambridge, 1999.

M. Aoun, R. Malti, F. Levron and A. Oustaloup, Numerical simulations of fractional systems: An overview of existing methods and improvements, Nonlinear Dynam. **38** (2004), no. 1-4, 117–131.

T.M. Atanacković, M. Janev, S. Pilipović and D. Zorica, An expansion formula for fractional derivatives of variable order, Cent. Eur. J. Phys. **11** (2013), no. 10, 1350–1360.

T.M. Atanacković, S. Konjik, Lj. Oparnica and S. Pilipović, Generalized Hamilton's principle with fractional derivatives, J. Phys. A **43** (2010), no. 25, 255203, 12 pp.

T.M. Atanacković, S. Konjik and S. Pilipović, Variational problems with fractional derivatives: Euler-Lagrange equations, J. Phys. A **41** (2008), no. 9, 095201, 12 pp.

T.M. Atanacković and B. Stanković, An expansion formula for fractional derivatives and its application, Fract. Calc. Appl. Anal. **7** (2004), no. 3, 365–378.

T.M. Atanacković and B. Stanković, On a numerical scheme for solving differential equations of fractional order, Mech. Res. Comm. **35** (2008), no. 7, 429–438.

F.M. Atici and P.W. Eloe, A transform method in discrete fractional calculus, Int. J. Difference Equ. **2** (2007), no. 2, 165–176.

F.M. Atici and P.W. Eloe, Initial value problems in discrete fractional calculus, Proc. Amer. Math. Soc. **137** (2009), no. 3, 981–989.

B. Aulbach and S. Hilger, A unified approach to continuous and discrete dynamics, in *Qualitative theory of differential equations (Szeged, 1988)*, 37–56, Colloq. Math. Soc. János Bolyai, 53 North-Holland, Amsterdam, 1990.

S.N. Avvakumov and Yu. N. Kiselev, Boundary value problem for ordinary differential equations with applications to optimal control, in *Spectral and evolution problems, Vol. 10 (Sevastopol, 1999)*, 147–155, Natl. Taurida University "V. Vernadsky", Simferopol, 2000.

D. Baleanu, O. Defterli and O.P. Agrawal, A central difference numerical scheme for fractional optimal control problems, J. Vib. Control **15** (2009), no. 4, 583–597.

D. Baleanu, K. Diethelm, E. Scalas and J.J. Trujillo, *Fractional calculus*, Series on Complexity, Nonlinearity and Chaos, 3, World Scientific Publishing Co. Pte. Ltd., Hackensack, NJ, 2012.

D. Baleanu and F. Jarad, Difference discrete variational principles, in *Mathematical analysis and applications*, 20–29, AIP Conf. Proc., 835 Amer. Inst. Phys., Melville, NY, 2006.

Z. Bartosiewicz and D.F.M. Torres, Noether's theorem on time scales, J. Math. Anal. Appl. **342** (2008), no. 2, 1220–1226.

N.R.O. Bastos, Fractional calculus on time scales, PhD thesis (Supervisor: D.F.M. Torres), Doctoral Programme in Mathematics and Applications (PDMA), University of Aveiro and University of Minho, 2012.

N.R.O. Bastos, R.A.C. Ferreira and D.F.M. Torres, Necessary optimality conditions for fractional difference problems of the calculus of variations, Discrete Contin. Dyn. Syst. **29** (2011a), no. 2, 417–437.

N.R.O. Bastos, R.A.C. Ferreira and D.F.M. Torres, Discrete-time fractional variational problems, Signal Processing **91** (2011b), no. 3, 513–524.

N.R.O. Bastos, D. Mozyrska and D.F.M. Torres, Fractional derivatives and integrals on time scales via the inverse generalized Laplace transform, Int. J. Math. Comput. **11** (2011), J11, 1–9.

N.R.O. Bastos and D.F.M. Torres, Combined Delta-Nabla Sum Operator in Discrete Fractional Calculus, Commun. Frac. Calc. 1 (2010), no. 1, 41–47.

R. Beals and R. Wong, *Special functions*, Cambridge Studies in Advanced Mathematics, 126, Cambridge University Press, Cambridge, 2010.

V.L. Berdichevsky, *Variational principles of continuum mechanics. II*, Interaction of Mechanics and Mathematics, Springer, Berlin, 2009.

S. Bhalekar, V. Daftardar-Gejji, D. Baleanu and R. Magin, Generalized fractional order bloch equation with extended delay, Internat. J. Bifur. Chaos Appl. Sci. Engrg. **22** (2012a), no. 4, 1250071, 15 pp.

S. Bhalekar, V. Daftardar-Gejji, D. Baleanu and R. Magin, Transient chaos in fractional Bloch equations, Comput. Math. Appl. **64** (2012b), no. 10, 3367–3376.

M. Bohner, Calculus of variations on time scales, Dynam. Systems Appl. **13** (2004), no. 3-4, 339–349.

M.J. Bohner, R.A.C. Ferreira and D.F.M. Torres, Integral inequalities and their applications to the calculus of variations on time scales, Math. Inequal. Appl. **13** (2010), no. 3, 511–522.

M. Bohner and A. Peterson, *Dynamic equations on time scales: an introduction with applications*, Birkhäuser, Boston, MA, 2001.

M. Bohner and A. Peterson, *Advances in dynamic equations on time scales*, Birkhäuser Boston, Boston, MA, 2003.

A.M.C. Brito da Cruz, N. Martins and D.F.M. Torres, Higher-order Hahn's quantum variational calculus, Nonlinear Anal. **75** (2012), no. 3, 1147–1157.

W.A. Brock, On existence of weakly maximal programmes in a multi-sector economy, Rev. Econ. Stud. **37** (1970), 275–280.

A.E. Bryson, Jr. and Y.C. Ho, *Applied optimal control*, Hemisphere Publishing Corp. Washington, DC, 1975.

P.L. Butzer, A.A. Kilbas and J.J. Trujillo, Mellin transform analysis and integration by parts for Hadamard-type fractional integrals, J. Math. Anal. Appl. **270** (2002), no. 1, 1–15.

P.L. Butzer, A.A. Kilbas and J.J. Trujillo, Stirling functions of the second kind in the setting of difference and fractional calculus, Numer. Funct. Anal. Optim. **24** (2003), no. 7-8, 673–711.

C. Canuto, M.Y. Hussaini, A. Quarteroni and T.A. Zang, *Spectral methods*, Scientific Computation, Springer, Berlin, 2007.

M. Caputo, Linear model of dissipation whose Q is almost frequency independent-II, Geophys. J. R. Astr. Soc. **13** (1967), 529–539.

F. Chen, X. Luo and Y. Zhou, Existence results for nonlinear fractional difference equation, Adv. Difference Equ. **2011** (2011), Art. ID 713201, 12 pp.

A.C. Chiang, *Elements of Dynamic Optimization*, McGraw-Hill Inc., Singapore, 1992.

C.F.M. Coimbra, Mechanics with variable-order differential operators, Ann. Phys. (8) **12** (2003), no. 11-12, 692–703.

C.F.M. Coimbra, C.M. Soon and M.H. Kobayashi, The variable viscoelasticity operator, Annalen der Physik **14** (2005), 378–389.

J. Cresson, G.S.F. Frederico and D.F.M. Torres, Constants of motion for non-differentiable quantum variational problems, Topol. Methods Nonlinear Anal. **33** (2009), no. 2, 217–231.

S. Das, *Functional fractional calculus for system identification and controls*, Springer, Berlin, 2008.

A. Debbouche and D.F.M. Torres, Approximate controllability of fractional non-local delay semilinear systems in Hilbert spaces, Internat. J. Control **86** (2013), no. 9, 1577–1585.

G. Diaz and C.F.M. Coimbra, Nonlinear dynamics and control of a variable order oscillator with application to the van der Pol equation, Nonlinear Dynam. **56** (2009), no. 1-2, 145–157.

J.B. Díaz and T.J. Osler, Differences of fractional order, Math. Comp. **28** (1974), 185–202.

K. Diethelm, *The analysis of fractional differential equations*, Lecture Notes in Mathematics, 2004, Springer, Berlin, 2010.

K. Diethelm, A fractional calculus based model for the simulation of an outbreak of dengue fever, Nonlinear Dyn. **71** (2013), no. 4, 613–619.

K. Diethelm, N.J. Ford and A.D. Freed, A predictor-corrector approach for the numerical solution of fractional differential equations, Nonlinear Dynam. **29** (2002), no. 1-4, 3–22.

K. Diethelm, N.J. Ford, A.D. Freed and Yu. Luchko, Algorithms for the fractional calculus: a selection of numerical methods, Comput. Methods Appl. Mech. Engrg. **194** (2005), no. 6-8, 743–773.

V.D. Djordjevic and T.M. Atanacković, Similarity solutions to nonlinear heat conduction and Burgers/Korteweg-de Vries fractional equations, J. Comput. Appl. Math. **222** (2008), no. 2, 701–714.

M. Dryl, A.B. Malinowska and D.F.M. Torres, A time-scale variational approach to inflation, unemployment and social loss, Control Cybernet. **42** (2013), no. 2, 399–418.

M. Dryl and D.F.M. Torres, The delta-nabla calculus of variations for composition functionals on time scales, Int. J. Difference Equ. **8** (2013), no. 1, 27–47.

M. Dryl and D.F.M. Torres, Necessary condition for an Euler-Lagrange equation on time scales, Abstr. Appl. Anal. **2014** (2014), Art. ID 631281, 7 pp.

M.Ö. Efe, Battery power loss compensated fractional order sliding mode control of a quadrotor UAV, Asian J. Control **14** (2012), no. 2, 413–425.

R.A. El-Nabulsi and D.F.M. Torres, Necessary optimality conditions for fractional action-like integrals of variational calculus with Riemann-Liouville derivatives of order (α, β), Math. Methods Appl. Sci. **30** (2007), no. 15, 1931–1939.

R.A. El-Nabulsi and D.F.M. Torres, Fractional actionlike variational problems, J. Math. Phys. **49** (2008), no. 5, 053521, 7 pp.

L. Elsgolts, *Differential equations and the calculus of variations*, translated from the Russian by George Yankovsky, Mir Publishers, Moscow, 1973.

S. Esmaeili and M. Shamsi, A pseudo-spectral scheme for the approximate solution of a family of fractional differential equations, Commun. Nonlinear Sci. Numer. Simul. **16** (2011), no. 9, 3646–3654.

R.A.C. Ferreira, Calculus of variations on time scales and discrete fractional calculus, PhD thesis (Supervisor: D.F.M. Torres; Co-supervisor: M. Bohner), University of Aveiro, 2010.

R.A.C. Ferreira and D.F.M. Torres, Higher-order calculus of variations on time scales, in *Mathematical control theory and finance* (eds: A. Sarychev, A. Shiryaev, M. Guerra, and M. do R. Grossinho), 149–159, Springer, Berlin, 2008.

R.A.C. Ferreira and D.F.M. Torres, Isoperimetric problems of the calculus of variations on time scales, in *Nonlinear Analysis and Optimization II* (eds: A. Leizarowitz, B. S. Mordukhovich, I. Shafrir, and A. J. Zaslavski), Contemporary Mathematics, vol. 514, Amer. Math. Soc., Providence, RI, 2010, 123–131.

R.A.C. Ferreira and D.F.M. Torres, Fractional h-difference equations arising from the calculus of variations, Appl. Anal. Discrete Math. **5** (2011), no. 1, 110–121.

N.J. Ford and J.A. Connolly, Comparison of numerical methods for fractional differential equations, Commun. Pure Appl. Anal. **5** (2006), no. 2, 289–306.

T. Fort, The calculus of variations applied to Nörlund's sum, Bull. Amer. Math. Soc. **43** (1937), no. 12, 885–887.

C.G. Fraser, Isoperimetric problems in the variational calculus of Euler and Lagrange, Historia Math. **19** (1992), no. 1, 4–23.

G.S.F. Frederico and D.F.M. Torres, A formulation of Noether's theorem for fractional problems of the calculus of variations, J. Math. Anal. Appl. **334** (2007), no. 2, 834–846.

G.S.F. Frederico and D.F.M. Torres, Fractional conservation laws in optimal control theory, Nonlinear Dynam. **53** (2008a), no. 3, 215–222.

G.S.F. Frederico and D.F.M. Torres, Fractional optimal control in the sense of Caputo and the fractional Noether's theorem, Int. Math. Forum **3** (2008b), no. 10, 479–493.

G.S.F. Frederico and D.F.M. Torres, Fractional Noether's theorem in the Riesz-Caputo sense, Appl. Math. Comput. **217** (2010), no. 3, 1023–1033.

W. Gautschi, *Orthogonal polynomials: computation and approximation*, Numerical Mathematics and Scientific Computation, Oxford University Press, New York, 2004.

H.H. Goldstine, *A history of the calculus of variations from the 17th through the 19th century*, Studies in the History of Mathematics and Physical Sciences, 5, Springer, New York, 1980.

C.S. Goodrich, Existence of a positive solution to a class of fractional differential equations, Appl. Math. Lett. **23** (2010a), no. 9, 1050–1055.

C.S. Goodrich, Continuity of solutions to discrete fractional initial value problems, Comput. Math. Appl. **59** (2010b), no. 11, 3489–3499.

P.D.F. Gouveia, A. Plakhov and D.F.M. Torres, Two-dimensional body of maximum mean resistance, Appl. Math. Comput. **215** (2009), no. 1, 37–52.

J. Gregory, Generalizing variational theory to include the indefinite integral, higher derivatives, and a variety of means as cost variables, Methods Appl. Anal. **15** (2008), no. 4, 427–435.

J. Gregory and C. Lin, Discrete variable methods for the m-dependent variable nonlinear, extremal problem in the calculus of variations. II, SIAM J. Numer. Anal. **30** (1993), no. 3, 871–881.

J. Gregory and R.S. Wang, Discrete variable methods for the m-dependent variable, nonlinear extremal problem in the calculus of variations, SIAM J. Numer. Anal. **27** (1990), no. 2, 470–487.

J. Hadamard, Essai sur l'étude des fonctions données par leur développement de Taylor, Journ. de Math. **4** (1892), no. 8, 101–186.

S. Hilger, Ein Maßkettenkalkül mit Anwendung auf Zentrumsmannigfaltigkeiten, PhD Thesis, Universität Würzburg, 1988.

S. Hilger, Analysis on measure chains—a unified approach to continuous and discrete calculus, Results Math. **18** (1990), no. 1-2, 18–56.

S. Hilger, Differential and difference calculus—unified!, Nonlinear Anal. **30** (1997), no. 5, 2683–2694.

R. Hilscher and V. Zeidan, Nonnegativity of a discrete quadratic functional in terms of the (strengthened) Legendre and Jacobi conditions, Comput. Math. Appl. **45** (2003), no. 6-9, 1369–1383.

R. Hilscher and V. Zeidan, Calculus of variations on time scales: weak local piecewise C_{rd}^1 solutions with variable endpoints, J. Math. Anal. Appl. **289** (2004), no. 1, 143–166.

R. Hilscher and V. Zeidan, Nonnegativity and positivity of quadratic functionals in discrete calculus of variations: survey, J. Difference Equ. Appl. **11** (2005), no. 9, 857–875.

M. Holm, The theory of discrete fractional calculus: Development and application, PhD thesis, University of Nebraska, 2010.

F. Jarad, T. Abdeljawad and D. Baleanu, Stability of q-fractional non-autonomous systems, Nonlinear Anal. Real World Appl. **14** (2013), no. 1, 780–784.

F. Jarad and D. Baleanu, Discrete variational principles for Lagrangians linear in velocities, Rep. Math. Phys. **59** (2007), no. 1, 33–43.

F. Jarad, D. Baleanu and T. Maraaba, Hamiltonian formulation of singular Lagrangians on time scales, Chinese Phys. Lett. **25** (2008), no. 5, 1720–1723.

Z.D. Jelicic and N. Petrovacki, Optimality conditions and a solution scheme for fractional optimal control problems, Struct. Multidiscip. Optim. **38** (2009), no. 6, 571–581.

V. Kac, P. Cheung, *Quantum Calculus*, Springer-Verlag, New York, 2002.

R. Kamocki, Pontryagin maximum principle for fractional ordinary optimal control problems, Math. Methods Appl. Sci. **37** (2014), no. 11, 1668–1686.

U.N. Katugampola, New approach to a generalized fractional integral, Appl. Math. Comput. **218** (2011), no. 3, 860–865.

W.G. Kelley and A.C. Peterson, *Difference equations*, Academic Press, Boston, MA, 1991.

H. Khosravian-Arab and D.F.M. Torres, Uniform approximation of fractional derivatives and integrals with application to fractional differential equations, Nonlinear Stud. **20** (2013), no. 4, 533–548.

A.A. Kilbas, Hadamard-type fractional calculus, J. Korean Math. Soc. **38** (2001), no. 6, 1191–1204.

A.A. Kilbas, H. M. Srivastava and J.J. Trujillo, *Theory and applications of fractional differential equations*, North-Holland Mathematics Studies, 204, Elsevier, Amsterdam, 2006.

A.A. Kilbas and A.A. Titioura, Nonlinear differential equations with Marchaud-Hadamard-type fractional derivative in the weighted space of summable functions, Math. Model. Anal. **12** (2007), no. 3, 343–356.

D.E. Kirk, *Optimal control theory: An introduction*, North-Holland Mathematics Studies, 204, Prentice-Hall Inc., Englewood Cliffs, NJ, 1970.

M. Klimek, Fractional sequential mechanics—models with symmetric fractional derivative, Czechoslovak J. Phys. **51** (2001), no. 12, 1348–1354.

M. Klimek and M. Lupa, Reflection symmetric formulation of generalized fractional variational calculus, Fract. Calc. Appl. Anal. **16** (2013), no. 1, 243–261.

P. Kumar and O.P. Agrawal, An approximate method for numerical solution of fractional differential equations, Signal Process. **86** (2006), no. 10, 2602–2610.

B. Kuttner, On differences of fractional order, Proc. London Math. Soc. (3) **7** (1957), 453–466.

V. Lakshmikantham, S. Sivasundaram and B. Kaymakcalan, *Dynamic systems on measure chains*, Mathematics and its Applications, 370, Kluwer Acad. Publ., Dordrecht, 1996.

M.J. Lazo and D.F.M. Torres, The DuBois-Reymond fundamental lemma of the fractional calculus of variations and an Euler-Lagrange equation involving only derivatives of Caputo, J. Optim. Theory Appl. **156** (2013), no. 1, 56–67.

M.J. Lazo and D.F.M. Torres, The Legendre condition of the fractional calculus of variations, Optimization **63** (2014), no. 8, 1157–1165.

L.P. Lebedev and M.J. Cloud, *The calculus of variations and functional analysis*, World Sci. Publishing, River Edge, NJ, 2003.

Y. Li, Y. Chen and H.-S. Ahn, Fractional-order iterative learning control for fractional-order linear systems, Asian J. Control **13** (2011), no. 1, 54–63.

Q. Lin, R. Loxton, K.L. Teo and Y.H. Wu, A new computational method for a class of free terminal time optimal control problems, Pac. J. Optim. **7** (2011), no. 1, 63–81.

Q. Lin, R. Loxton, K.L. Teo and Y.H. Wu, Optimal control computation for nonlinear systems with state-dependent stopping criteria, Automatica J. IFAC **48** (2012), no. 9, 2116–2129.

C.F. Lorenzo and T.T. Hartley, Variable order and distributed order fractional operators, Nonlinear Dynam. **29** (2002), no. 1-4, 57–98.

A. Lotfi, M. Dehghan and S.A. Yousefi, A numerical technique for solving fractional optimal control problems, Comput. Math. Appl. **62** (2011), no. 3, 1055–1067.

S. Ma, Y. Xu and W. Yue, Numerical solutions of a variable-order fractional financial system, J. Appl. Math. **2012** (2012), Art. ID 417942, 14 pp.

A.B. Malinowska and D.F.M. Torres, Strong minimizers of the calculus of variations on time scales and the Weierstrass condition, Proc. Est. Acad. Sci. **58** (2009), no. 4, 205–212.

A.B. Malinowska and D.F.M. Torres, Natural boundary conditions in the calculus of variations, Math. Methods Appl. Sci. **33** (2010a), no. 14, 1712–1722.

A.B. Malinowska and D.F.M. Torres, Leitmann's direct method of optimization for absolute extrema of certain problems of the calculus of variations on time scales, Appl. Math. Comput. **217** (2010b), no. 3, 1158–1162.

A.B. Malinowska and D.F.M. Torres, The Hahn quantum variational calculus, J. Optim. Theory Appl. **147** (2010c), no. 3, 419–442.

A.B. Malinowska and D.F.M. Torres, Delta-nabla isoperimetric problems, Int. J. Open Probl. Comput. Sci. Math. **3** (2010d), no. 4, 124–137.

A.B. Malinowska and D.F.M. Torres, Generalized natural boundary conditions for fractional variational problems in terms of the Caputo derivative, Comput. Math. Appl. **59** (2010e), no. 9, 3110–3116.

A.B. Malinowska and D.F.M. Torres, *Introduction to the fractional calculus of variations*, Imperial College Press, London, 2012.

A.B. Malinowska and D.F.M. Torres, *Quantum variational calculus*, Springer Briefs in Electrical and Computer Engineering, Springer, Cham, 2014.

N. Martins and D.F.M. Torres, Calculus of variations on time scales with nabla derivatives, Nonlinear Anal. **71** (2009), no. 12, e763–e773.

N. Martins and D.F.M. Torres, Generalizing the variational theory on time scales to include the delta indefinite integral, Comput. Math. Appl. **61** (2011), no. 9, 2424–2435.

M.M. Meerschaert and C. Tadjeran, Finite difference approximations for fractional advection-dispersion flow equations, J. Comput. Appl. Math. **172** (2004), no. 1, 65–77.

K.S. Miller and B. Ross, Fractional difference calculus, in *Univalent functions, fractional calculus, and their applications (Kōriyama, 1988)*, 139–152, Ellis Horwood Ser. Math. Appl, Horwood, Chichester, 1989.

K.S. Miller and B. Ross, *An introduction to the fractional calculus and fractional differential equations*, A Wiley-Interscience Publication, Wiley, New York, 1993.

D. Mozyrska, Multiparameter fractional difference linear control systems, Discrete Dyn. Nat. Soc. **2014** (2014), Art. ID 183782, 8 pp.

D. Mozyrska and E. Girejko, Overview of fractional *h*-difference operators, in *Advances in harmonic analysis and operator theory*, 253–268, Oper. Theory Adv. Appl., 229, Birkhäuser/Springer Basel AG, Basel, 2013.

D. Mozyrska and D.F.M. Torres, Minimal modified energy control for fractional linear control systems with the Caputo derivative, Carpathian J. Math. **26** (2010), no. 2, 210–221.

D. Mozyrska and D.F.M. Torres, Modified optimal energy and initial memory of fractional continuous-time linear systems, Signal Process. **91** (2011), no. 3, 379–385.

T. Odzijewicz, A.B. Malinowska and D.F.M. Torres, Fractional variational calculus with classical and combined Caputo derivatives, Nonlinear Anal. **75** (2012a), no. 3, 1507–1515.

T. Odzijewicz, A.B. Malinowska and D.F.M. Torres, Fractional calculus of variations in terms of a generalized fractional integral with applications to physics, Abstr. Appl. Anal. **2012** (2012b), Art. ID 871912, 24 pp.

T. Odzijewicz, A.B. Malinowska and D.F.M. Torres, Variable order fractional variational calculus for double integrals, Proceedings of the 51st IEEE Conference on Decision and Control, December 10–13, 2012, Maui, Hawaii, Art. no. 6426489 (2012c), 6873–6878.

T. Odzijewicz, A.B. Malinowska and D.F.M. Torres, Fractional variational calculus of variable order, Advances in Harmonic Analysis and Operator Theory, The Stefan Samko Anniversary Volume (Eds: A. Almeida, L. Castro, F.-O. Speck), Operator Theory: Advances and Applications **229** (2013a), Springer, 291–301.

T. Odzijewicz, A.B. Malinowska and D.F.M. Torres, Noether's theorem for fractional variational problems of variable order, Cent. Eur. J. Phys. **11** (2013b), no. 6, 691–701.

T. Odzijewicz, A.B. Malinowska and D.F.M. Torres, A generalized fractional calculus of variations, Control Cybernet. **42** (2013c), no. 2, 443–458.

T. Odzijewicz and D.F.M. Torres, Fractional calculus of variations for double integrals, Balkan J. Geom. Appl. **16** (2011), no. 2, 102–113.

M.D. Ortigueira, The fractional quantum derivative and its integral representations, Commun. Nonlinear Sci. Numer. Simul. **15** (2010a), no. 4, 956–962.

M.D. Ortigueira, On the fractional linear scale invariant systems, IEEE Trans. Signal Process. **58** (2010b), no. 12, 6406–6410.

M.D. Ortigueira, *Fractional calculus for scientists and engineers*, Lecture Notes in Electrical Engineering, 84, Springer, Dordrecht, 2011.

M.D. Ortigueira and F.V. Coito, The initial conditions of Riemann-Liouville and Caputo derivatives, in: Proceedings of the Sixth EUROMECH Conference ENOC 2008, June 30-July 4, 2008, Saint Petersburg, Russia.

M.D. Ortigueira and F.V. Coito, On the usefulness of Riemann-Liouville and Caputo derivatives in describing fractional shift-invariant linear systems, Journal of Applied Nonlinear Dynamics **1** (2012), no. 2, 113–124.

A. Yu. Plakhov and D.F.M. Torres, Newton's aerodynamic problem in a medium consisting of moving particles, Mat. Sb. **196** (2005), no. 6, 111–160; translation in Sb. Math. **196** (2005), no. 5-6, 885–933.

I. Podlubny, *Fractional differential equations*, Mathematics in Science and Engineering, 198, Academic Press, San Diego, CA, 1999.

L.S. Pontryagin, V.G. Boltyanskii, R.V. Gamkrelidze and E.F. Mishchenko, *The mathematical theory of optimal processes*, Translated from the Russian by K.N. Trirogoff; edited by L.W. Neustadt, Interscience Publishers John Wiley & Sons Inc., New York, 1962.

S. Pooseh, R. Almeida and D.F.M. Torres, Approximation of fractional integrals by means of derivatives, Comput. Math. Appl. **64** (2012a), no. 10, 3090–3100.

S. Pooseh, R. Almeida and D.F.M. Torres, Discrete direct methods in the fractional calculus of variations, Proceedings of FDA'2012, May 14-17, 2012, Hohai University, Nanjing, China, 2012b. Paper #042.

S. Pooseh, R. Almeida and D.F.M. Torres, Expansion formulas in terms of integer-order derivatives for the Hadamard fractional integral and derivative, Numer. Funct. Anal. Optim. **33** (2012c), no. 3, 301–319.

S. Pooseh, R. Almeida and D.F.M. Torres, Numerical approximations of fractional derivatives with applications, Asian J. Control **15** (2013a), no. 3, 698–712.

S. Pooseh, R. Almeida and D.F.M. Torres, Discrete direct methods in the fractional calculus of variations, Comput. Math. Appl. **66** (2013b), no. 5, 668–676.

S. Pooseh, R. Almeida and D.F.M. Torres, Free time fractional optimal control problems, Proceedings of the European Control Conference 2013, Zurich, Switzerland, July 17-19, 2013c, 3985–3990.

S. Pooseh, R. Almeida and D.F.M. Torres, A numerical scheme to solve fractional optimal control problems, Conference Papers in Mathematics **2013** (2013d), Art. ID 165298, 10 pp.

S. Pooseh, R. Almeida and D.F.M. Torres, A discrete time method to the first variation of fractional order variational functionals, Cent. Eur. J. Phys. **11** (2013e), no. 10, 1262–1267.

S. Pooseh, R. Almeida and D.F.M. Torres, Fractional order optimal control problems with free terminal time, J. Ind. Manag. Optim. **10** (2014), no. 2, 363–381.

S. Pooseh, H.S. Rodrigues and D.F.M. Torres, Fractional derivatives in dengue epidemics, Numerical Analysis and Applied Mathematics ICNAAM 2011, AIP Conf. Proc. **1389** (2011), no. 1, 739–742.

D. Qian, Z. Gong and C. Li, A generalized Gronwall inequality and its application to fractional differential equations with Hadamard derivatives, 3rd Conference on Nonlinear Science and Complexity (NSC10), Cankaya University, 1–4, Ankara, Turkey, 28–31 July, 2010.

E.M. Rabei, K.I. Nawafleha, R.S. Hijjawia, S.I. Muslihc and D. Baleanu, The Hamilton formalism with fractional derivatives, J. Math. Anal. Appl. **327** (2007), no. 2, 891–897.

L.E.S. Ramirez and C.F.M. Coimbra, On the selection and meaning of variable order operators for dynamic modeling, Int. J. Differ. Equ. **2010** (2010), Art. ID 846107, 16 pp.

L.E.S. Ramirez and C.F.M. Coimbra, On the variable order dynamics of the nonlinear wake caused by a sedimenting particle, Phys. D **240** (2011), no. 13, 1111–1118.

A. Razminia and D.F.M. Torres, Control of a novel chaotic fractional order system using a state feedback technique, Mechatronics **23** (2013), no. 7, 755–763.

F. Riewe, Nonconservative Lagrangian and Hamiltonian mechanics, Phys. Rev. E (3) **53** (1996), no. 2, 1890–1899.

F. Riewe, Mechanics with fractional derivatives, Phys. Rev. E (3) **55** (1997), no. 3, part B, 3581–3592.

H.S. Rodrigues, M.T.T. Monteiro and D.F.M. Torres, Dynamics of Dengue epidemics when using optimal control, Math. Comput. Modelling **52** (2010a), 1667–1673.

H.S. Rodrigues, M.T.T. Monteiro and D.F.M. Torres, Insecticide control in a Dengue epidemics model, Numerical Analysis and Applied Mathematics, edited by T. Simos, AIP Conf. Proc. **1281** (2010b), no. 1, 979–982.

H.S. Rodrigues, M.T.T. Monteiro, D.F.M. Torres and A. Zinober, Dengue disease, basic reproduction number and control, Int. J. Comput. Math. **89** (2012), no. 3, 334–346.

B. Ross, S.G. Samko and E.R. Love, Functions that have no first order derivative might have fractional derivatives of all orders less than one, Real Anal. Exchange **20** (1994), 140–157.

S.G. Samko, Fractional integration and differentiation of variable order, Anal. Math. **21** (1995), no. 3, 213–236.

S.G. Samko, A.A. Kilbas and O.I. Marichev, *Fractional integrals and derivatives*, translated from the 1987 Russian original, Gordon and Breach, Yverdon, 1993.

S.G. Samko and B. Ross, Integration and differentiation to a variable fractional order, Integral Transform. Spec. Funct. **1** (1993), no. 4, 277–300.

S. Sengul, Discrete Fractional Calculus and Its Applications to Tumor Growth, Master Thesis, 2010.

J. Shen and J. Cao, Necessary and sufficient conditions for consensus of delayed fractional-order systems, Asian J. Control **14** (2012), no. 6, 1690–1697.

M.R. Sidi Ammi, E. H. El Kinani and D.F.M. Torres, Existence and uniqueness of solutions to functional integro-differential fractional equations, Electron. J. Differential Equations **2012** (2012), no. 103, 9 pp.

M.R. Sidi Ammi and D.F.M. Torres, Existence and uniqueness of a positive solution to generalized nonlocal thermistor problems with fractional-order derivatives, Differ. Equ. Appl. **4** (2012), no. 2, 267–276.

C.J. Silva and D.F.M. Torres, Two-dimensional Newton's problem of minimal resistance, Control Cybernet. **35** (2006), no. 4, 965–975.

V. Spedding, Taming nature's numbers, New Scientist, 19 July 2003, 28–31.

J. Stoer and R. Bulirsch, *Introduction to numerical analysis*, translated from the German by R. Bartels, W. Gautschi and C. Witzgall, third edition, Texts in Applied Mathematics, 12, Springer, New York, 2002.

W. Sun, Y. Li, C. Li and Y. Chen, Convergence speed of a fractional order consensus algorithm over undirected scale-free networks, Asian J. Control **13** (2011), no. 6, 936–946.

H.J. Sussmann and J.C. Willems, The brachistochrone problem and modern control theory, in *Contemporary trends in nonlinear geometric control theory and its applications (México City, 2000)*, 113–166, World Sci. Publ., River Edge, NJ, 2002.

J.A. Tenreiro Machado, V. Kiryakova and F. Mainardi, Recent history of fractional calculus, Commun. Nonlinear Sci. Numer. Simul. **16** (2011), no. 3, 1140–1153.

J.A. Tenreiro Machado, M.F. Silva, R.S. Barbosa, I.S. Jesus, C.M. Reis, M.G. Marcos and A.F. Galhano, Some applications of fractional calculus in engineering, Math. Probl. Eng., Art. ID 639801 (2010), 34 pp.

D.F.M. Torres, The variational calculus on time scales, Int. J. Simul. Multidisci. Des. Optim. **4** (2010), no. 1, 11–25.

L.N. Trefethen, *Spectral methods in MATLAB*, Software, Environments, and Tools, 10, SIAM, Philadelphia, PA, 2000.

C. Tricaud and Y. Chen, An approximate method for numerically solving fractional order optimal control problems of general form, Comput. Math. Appl. **59** (2010a), no. 5, 1644–1655.

C. Tricaud and Y. Chen, Time-optimal control of systems with fractional dynamics, Int. J. Differ. Equ. **2010** (2010b), Art. ID 461048, 16 pp.

J.C. Trigeassou, N. Maamri and A. Oustaloup, The infinite state approach: Origin and necessity, Comput. Math. Appl. **66** (2013), no. 5, 892–907.

C. Tuckey, *Nonstandard methods in the calculus of variations*, Pitman Research Notes in Mathematics Series, 297, Longman Sci. Tech., Harlow, 1993.

D. Valério and J.S. Costa, Variable-order fractional derivatives and their numerical approximations, Signal Process. **91** (2011), no. 3, 470–483.

B. van Brunt, *The calculus of variations*, Universitext, Springer, New York, 2004.

S.A. Yousefi, A. Lotfi and M. Dehghan, The use of a Legendre multiwavelet collocation method for solving the fractional optimal control problems, J. Vib. Control **17** (2011), no. 13, 2059–2065.

P. Zhuang, F. Liu, V. Anh and I. Turner, Numerical methods for the variable-order fractional advection-diffusion equation with a nonlinear source term, SIAM J. Numer. Anal. **47** (2009), no. 3, 1760–1781.

Index

Δ-integral, 184

Antiderivative, 184

Backward graininess function, 181
Backward jump operator, 181
Binomial
  coefficient, 24
  theorem, 64, 99
Boundary conditions, 12, 32
Brachistochrone, 10

Caputo fractional derivative
  left, 27, 63
  right, 27, 63

Direct method, 15, 17, 49, 111
Dubois–Reymond lemma on time
  scales, 187

Error
  function, 104
  truncation, 65, 81, 89, 108
Euler method, 15, 112
Euler–Lagrange equation, 13, 14, 132
  fractional, 32, 114, 118, 129
  discrete fractional case, 198
  time scales case, 188
Expansion formula, 57, 59

Finite
  differences, 13, 16, 24, 49, 111

element, 53
Forward graininess function, 181
Forward jump operator, 181
Fractional
  differential equation, 31, 72, 78,
    101, 122
  dynamics, 30
  integral equation, 96, 97, 101
  optimal control, 30
Fractional calculus of variable order,
  160
Fractional summation by parts, 195
Function
  Δ-differentiable, 182
  rd-continuous, 183
Functional, 11
  augmented, 18
  differentiable, 11
  increment of, 11
Fundamental
  lemma, 14, 33
  theorem, 11

Gamma function, 21
Gateaux derivative, 14, 33
Grünwald–Letnikov fractional
  derivative
  left, 25, 111
  right, 25

Hadamard fractional derivative
  left, 26, 74, 76, 83

right, 26, 74, 77
Hadamard fractional integral
    left, 26, 101
    right, 26, 103
Hamiltonian, 18, 130, 135, 140, 142, 149
Hamiltonian system, 142, 148
Hypergeometric function, 64, 88

Indirect method, 15, 17, 50, 129
Integration by parts, 13, 86
    fractional, 28, 33
Isoperimetric problem, 12, 45, 124

Kantorovich method, 15

Lagrange multipliers, 31, 51, 130, 134
Lagrangian, 12, 116, 117
Left fractional difference, 194, 221
Left fractional sum, 190
Legendre's necessary condition
    time scales case, 188

MATLAB, 54, 74, 153, 155, 243
Minimum Principle, 18, 50
Mittag–Leffler function, 22, 67
Moment of a function, 59, 60, 86, 97, 134

Optimal
    control, 17
    solution, 32
Optimal time problem, 47

Points
    left-dense, 181
    left-scattered, 181
    right-dense, 181
    right-scattered, 181
Polynomials on time scales, 185

Queen Dido, 9

Regressive
    equation, 213
Regulated function, 181

Riemann–Liouville fractional
    derivative
    left, 23, 29, 57, 59, 83, 111
    right, 23, 59–61, 66
Riemann–Liouville fractional integral
    left, 23, 85
    right, 23, 89
Riesz fractional derivative, 42
Riesz fractional integral, 42
Riesz–Caputo fractional derivative, 42
Right fractional difference, 194, 221
Right fractional sum, 190
Ritz method, 15, 52

Shifted Legendre polynomials, 52
Stationary condition, 142, 148

Time Scale, 180
Transversality conditions, 142, 143, 149

Variation, 11, 114, 122, 123, 140
    of a functional, 13, 33
Variational problem, 12, 17, 49
    fractional, 29, 31

Weakly maximizer, 34